Audel™ House Wiring

All New 8th Edition

Paul Rosenberg
Roland Palmquist

Wiley Publishing, Inc.

Vice President and Executive Group Publisher: Richard Swadley
Vice President and Executive Publisher: Bob Ipsen
Vice President and Publisher: Joseph B. Wikert
Executive Editorial Director: Mary Bednarek
Editorial Manager: Kathryn A. Malm
Executive Editor: Carol A. Long
Production Manager: Gerry T. Fahey
Development Editor: Eileen Bien Calabro
Production Editor: Ava Wilder
Text Design & Composition: TechBooks

Published simultaneously in Canada

For general information on our other products and services please contact our Customer Care Department within the United States at (800) 762-2974, outside the United States at (317) 572-3993 or fax (317) 572-4002.

Wiley also publishes its books in a variety of electronic formats. Some content that appears in print may not be available in electronic books.

Library of Congress Cataloging-in-Publication Data:

ISBN: 0-764-56956-2

Printed in the United States of America

10 9 8 7 6 5 4 3 2 1

Contents

Foreword

It has been a pleasure for me to put together a house wiring text that would provide serious trade knowledge to a general audience. It has been my experience that virtually all do-it-yourself books are short on seriousness and long on "five easy steps."

Installing electrical wiring is not supposed to be simple or easy. Electrical wiring is a serious undertaking. Basic wiring projects are not beyond the skill of a homeowner, but neither are they something that can be taken lightly. I do not like treating wiring as if it were as simple as applying paint to a wall; it is not. Doable? Yes, definitely. But never "quick and easy."

In this book, I have taken the information that is really necessary to wire a house properly, and have put it into an order and form that can be understood by a nonprofessional. This is not a dumbed-down book. It is a professional level text, with a bit of extra explanation added. Furthermore, this book covers only house wiring. Narrowing the subject allows for a book that covers the material fully without being overwhelmingly long or complex.

- Chapter 1 covers the basics of the *National Electrical Code* (*NEC*) and general safety. The *NEC* contains all of the rules that are necessary to perform safe electrical installations. They are also the rules that all electrical installations are required to meet. Electrical wiring that does not meet *NEC* requirements will not be approved and is probably unsafe. So, understanding the *NEC* is absolutely critical for any installer of electrical wiring. This chapter explains what the *NEC* is all about, the particular wording it uses, and how to use it with a minimum of difficulty.
- Chapter 2 details the requirements and methods for installing an electrical service entrance. This is the part of an electrical system that is exposed to the most hazards, and it needs its own careful explanation.
- Chapter 3 deals with the branch circuits that power all of the receptacles, lights, and appliances we use in our homes.
- Chapter 4 covers the materials and methods for house wiring—the various cables, wires, conduits, and boxes that make up a complete electrical installation. All of these are items that must be applied properly. Again, the rules and methods are not especially difficult, but they do need to be addressed directly.

- Chapter 5 covers electric heating systems, which are very important in some areas and seldom used in others.
- Chapter 6 covers mobile homes, which have a few special requirements and must be wired correctly to avoid hazards.
- Chapter 7 covers several types of communication wiring (telephone, cable TV, broadband Internet, and home networks) as well as security systems. These systems are not generally as hazardous as wiring for light and power, but they must be installed correctly if they are to do their jobs. Installing them is in many ways easier than installing power wiring.
- Chapter 8 is unique in that it is especially focused on the process of actually wiring a house. All of the text in this chapter follows the process of wiring a house, roughly from beginning to end.
- Appendix A explains and gives examples of load calculations. These calculations are frequently required to obtain an electrical permit. This Appendix outlines the process step-by-step.
- Appendix B covers the requirements and calculations for farm buildings.
- Appendix C covers the requirements for installation of audio and home theater systems.
- Appendix D discusses the components and connections needed for various methods of delivering electronic entertainment obtained from the Internet.

Whether professional electrician or homeowner, I think that you'll find that this book contains all of the essential information you need to wire a home properly. Read it carefully, and have it nearby as you undertake your house-wiring project.

Paul Rosenberg

Chapter 1

Wiring Basics

In any complex undertaking there are fundamental issues that must be mastered if the project is to be successful. For the wiring of a house, those fundamentals are the safety of the installation and the safety of the installer. Everything else follows from there.

The wiring you install must not hurt people or damage property. This wiring may remain in place and function for many decades, and it may be used by dozens if not hundreds of people. Your first concern is that it operate safely. Through much of this book, we will explain how to build a house wiring system that will provide the services and capacities you need. Those requirements are certainly very important, but they are a far second in importance to safety. Wiring systems can always be expanded and improved, but an unsafe system cannot be tolerated.

As we proceed through this text, and as you proceed through your wiring project, you can never entirely let this thought out of your mind: Electric power is one of the most useful things humans have ever discovered and is a great blessing upon humanity, but it is an impersonal and amoral force. If wiring is not installed properly, it will cause harm just as readily as it will cause benefit.

Safety

As just mentioned, there are two primary safety concerns in house wiring: the safety of the installation and the safety of the installer. We will begin our discussion with the safety of the installer.

Construction Site Safety

Your first concern is job-site safety. We are discussing construction projects here, and construction sites are inherently hazardous. In fact, most accidents suffered by electricians are related to general construction hazards rather than electrical shock. Such hazards as stepping on nails or screws, air hammer accidents, loose flooring, falls, and falling construction materials are the most common causes of injuries.

When working on any construction site, awareness is your first priority. What are the carpenters doing? Who else has been to the job? What did they install? Did they complete it? Is it safe yet? You cannot assume anything on a construction site. Any long-time construction worker can tell you stories about people they knew who were hurt (or worse) simply because they assumed something.

Your eyes must be open at all times on any construction site. This is your job, and you cannot delegate it to anyone else—ever.

It is difficult to get this message across strongly enough. House wiring is simpler than industrial wiring, to be sure, but it still requires that the installer be informed, cautious, and sensible. Generally, young construction workers never understand this necessity completely until they have a near miss (or worse). Then they begin to understand. Please try to avoid that pattern.

As for avoiding electrical shock or burn, additional care must be taken to know which wire is connected to what, and to maintain color codes strictly. Hot wires in a house should always be black or red. Neutrals should always be white, grounds always green, and only switch leg conductors should be other colors. This discipline cannot be deviated from, or you will be in jeopardy.

The risks from wiring are shock and burn. Shocking generally occurs when your body (or part of it) gets into some position where it completes a circuit. That is, your body becomes a conductor—flesh substituting for a wire. Usually this happens when you touch a hot wire with one part of your body while another part of your body is touching ground. For house wiring at 120 volts to ground, this is painful but seldom leads to real damage. However, if you are very well grounded (standing in a puddle of water, for example), death is certainly possible. Be careful.

Burning generally occurs when you accidentally put a tool in between two live conductors, thus creating a short circuit. In these cases, a very high level of current will immediately begin to flow through your screwdriver, pliers, or whatever piece of metal you have put where it doesn't belong. Such high currents cause so much heating that they will, within milliseconds, melt and/or vaporize the copper of the conductors, the steel of the tool, or both. Because gas takes many, many times more volume than the same mass of a solid, an explosion results. The solid copper that occupied only one-tenth of a cubic centimeter now requires several cubic centimeters as a gas. And since this transition takes only a fraction of a second, it blows molten copper away from the location of the conductors and toward you (among other things). In most cases, at residential levels of current and voltage, such accidents result in burns that heal in a week or two. But worse can happen, and occasionally does. Again, be careful.

Remember that electricity can injure and kill people and ignite fires. Higher voltages are obviously more dangerous than lower voltages, but even normal household voltages can hurt you. If you are not experienced with electrical work, do not work on live circuits!

If you must test energized circuits, finish your work, make sure no one will contact the electrical system, turn on the power with a circuit breaker, test, then turn power off again. Working *hot* (that is, with circuits energized) is only for experienced electricians, and even they avoid it whenever possible. Don't take risks—it isn't worth it. If you are wiring your own home and come to a situation where you need to work hot, hire a local electrician for a couple of hours. It will be money well spent.

The Safety of the Installation

For the remained of this chapter we will explain how wiring installations are made safe for use. This is accomplished by adhering to the *National Electrical Code*. This document is a condensation of a hundred years of experience in electrical wiring. Professional electricians, engineers, and manufacturers maintain this document and modify it continually. Its purpose is to define rules for safe electrical installations. And after so long a time, it has become a very good set of rules. Don't second-guess these rules.

The *National Electrical Code* contains time-tested rules for the installation of electrical wiring. But these rules, however good, must be applied by an installer. That makes the installer the key element in the equation. The rules of the *National Electrical Code* are not overly difficult to understand and apply, or else there could not be hundreds of thousands of electrical workers operating in the trade every day. But applying these rules takes attention and thought. House wiring is not beyond your ability, but it will require you to think about what you are doing, not merely to go through the motions.

Finally, when doing any sort of construction work, be sure that you have proper tools. Get the same tools that full-time electricians use. They can be found at thousands of electrical wholesale companies. Using the proper tools is especially important for ladders, lifts, and scaffolds. Don't take risks.

The *National Electrical Code*

Virtually all home electrical installations must conform to the *National Electrical Code* (which we will frequently call the *NEC* or the *Code*), published by the National Fire Protection Association, Inc. (NFPA). The NFPA came into being around the beginning of the 20th century to regulate new uses of electricity and other types of energy such as fuel. Until then there had been no rules regarding the installation of electrical wiring. As the number of electrical wiring installations began to rise, so did incidence of fires caused by electricity. Because of this, a number of insurance companies set up a committee to write rules for safe electrical installation.

After these rules were published, the insurance companies agreed that none of them would insure structures whose electrical wiring did not conform to the new rules. Thus, the *NEC* became a set of rules that defined safe electrical installation. In the course of time, most of the enforcement of these rules was taken over by local building inspectors, but the net effect remains the same: If these rules are not followed, a building cannot be used.

The *Code* spells out specific requirements, but most of them are centered on commonsense concepts, such as the following:

- Don't allow any electrical device or component to overheat.
- Don't allow foreign substances to contact energized components.
- Don't allow any materials to be stressed beyond their tolerances.
- Protect electrical systems and devices from being overloaded, in the present or the future.
- Keep all energized parts protected, and avoid any sort of accidental contact.
- Use only high-quality materials that will not fail under expected uses and conditions.
- See that only qualified persons perform installations.
- Do not allow electrical installations to create a hazardous situation.
- Verify that critical systems are ultra-reliable.

Concepts such as these are the essence of the *NEC*. The goal is to provide electrical installations that do not injure people or harm property.

There is, however, a difficulty that arises when we try to apply common sense to complex systems: Common sense is relatively easy to apply to simple, well-known situations; it is much harder to use when the factors are too numerous, or too difficult to identify. It is for this reason that a document like the *NEC* must exist. It contains the collected knowledge of thousands of people, over a period of decades, as to what the critical factors are and how they are best addressed. No one of us has enough experience to identify all the necessary information on every type of electrical installation.

Therefore, a central depository of accumulated knowledge is required. That is what the *NEC* really is.

The engineers who write the *Code* are primarily concerned with correctness and completeness. They generally do a very good job, but that does not ensure that their work will be easily understood by the people who must use the *Code*. In fact, it is the very effort to make the *Code* complete and correct that also makes it difficult to understand.

A given article of the *NEC* may contain information that is pertinent to engineers only, other information pertinent to manufacturers only, and still more information of use only to installers. For example, some of the requirements for markings on cables are not important to an installer. Yet all of this information is lumped together into one brief article that must be dissected by the reader. This is where the difficulty most often lies.

Key Words

When reading and interpreting the *NEC*, you must pay special attention to certain words. These few words have broad implications and are given immense weight by any interpreter, such as an electrical inspector. You must pay careful attention for the following key words:

- **Shall:** Any time you see the word shall in the *NEC*, it means that you *must* do something in a certain way. You have *no* choice at all; either you do it that specific way, or you are in violation of the *Code*.
- **May:** The word may gives you an option. You can do it the specific way that is stated, or you can do it another way; it is your choice.
- **Grounded conductor:** This is almost always the neutral conductor. Take care not to let the word grounded confuse you. Grounded conductor does not refer to a green wire.
- **Grounding conductor:** This is the green (or bare) wire, more correctly called the *equipment-grounding conductor.* The grounding conductor is used to connect equipment to ground.

In addition to these terms, there are other, less common terms that can also be confusing. Remember that the *NEC* cannot be read casually. To make correct interpretations, you must consider each word individually. This requires extra time and more effort. There is, however, no other way to arrive at correct interpretations.

How the Code Is Arranged

The *NEC* is divided into nine chapters. *Chapters 1, 2, 3,* and *4* cover general requirements. They are the broadest, most basic rules. *Chapter 1* covers the most basic rules; *Chapter 2*, the basic requirements of grounding, circuits, services, and overcurrent protection; *Chapter 3*, wiring methods such as conduit and cables; and *Chapter 4*, equipment such as cords, lighting fixtures, switches and receptacles. In these sections, you will find the basic rules for all wiring installations.

Chapters 5, 6, and *7* apply to special occupancies, special equipment, or other special conditions. These sections either supplement or modify the rules of *Chapters 1* through *4*. In other words, *Chapters 1, 2, 3,* and *4* state rules that will always stand, except if modified specifically by a later section of the code.

Only specific exceptions, for specific types of installations and specific conditions, can change the rules of *Chapters 1* through *4*.

Chapter 8 covers communications systems and is independent of the other chapters except where they are specifically referenced therein, and *Chapter 9* consists of tables.

Three Types of Rules

There are types of rules that you will find listed in the *NEC*: mandatory rules, permissive rules, and explanatory material. *Mandatory rules* are characterized by the use of the words shall or shall not. *Permissive rules* are characterized by the use of the words shall be permitted or shall not be required. Material such as references to other standards, other sections of the *NEC*, or related information, is included in the form of *fine-print notes* (identified by small print and the abbreviation FPN), which are informational only and are not enforceable as requirements of this *Code*.

Section Numbers

As you progress with the reading of this book, you will find references to certain portions of the *Code*, such as *Section 230.70* or *Table 370.6(A)*. These indicate the sections or tables as numbered in the *NEC* so that, should you wish to do so, you may look them up in the *NEC* for further study. There will also be quotations from the *NEC* as well as some tables. The article and section numbers as well as quotes from them will be in italics, and the tables will be identified by notes.

General Requirements

The *NEC* is a large book, containing thousands of requirements. But there are a few that are more basic and more central to safety than the others. These are contained in *Section 110* of the *Code*, and it is important to understand them.

The most basic and important of these requirements is *Section 110.12*, which states that all installations must be performed "*in a neat and workmanlike manner.*" In other words, all electrical installation requirements presuppose that the installer is concerned, informed, and mindful. Without this prerequisite, all other requirements are almost valueless.

Other critical portions of this article are as follows:

- *110.5. Conductors.* All of the requirements in the *NEC* that regard conductors refer to copper conductors. That is the default. In cases where other types of conductors are permitted (as with aluminum conductors), they must have their sizes changed accordingly.
- *110.7. Insulation Integrity.* All finished wiring is required to be free from short circuits and from unintentional grounds.
- *110.9. Interrupting Rating.* Equipment intended to interrupt current (such as fuses or circuit breakers) must have an interrupting rating sufficient for the voltage and the current levels that are available to it. The device must be able to function properly in the circuit where it is installed.
- *110.11. Deteriorating Agents.* Only equipment that is identified for such use may be installed in a damp or wet location. This also applies to any location exposed to gases, fumes, vapors, liquids, or deteriorating agents or to excessive temperatures.
- *110.12. Mechanical Execution of Work.* After the "*neat and workmanlike manner*" portion that we explained earlier, this section goes on to say that unused openings in boxes, raceways, and the like must be effectively closed, that conductors in underground enclosures must be racked, and that the various parts of electrical equipment must be kept clean and undamaged.
- *110.13. Mounting and Cooling of Equipment.* Electrical equipment must be firmly mounted. Wooden plugs (an old method of mounting equipment on masonry) may not be used. Electrical equipment that depends on air circulation for cooling must not have its ventilation restricted.
- *110.14. Electrical Connections.* Every conductor termination device (such as a lug, a terminal, or a wire nut) must be

identified as to the types of conductors it is designed for, and it must be used in accordance with these markings. There is particular concern over the combining of dissimilar metals, such as copper and aluminum. These may never by connected together, except with a device specifically identified for that use. Splices must be made with devices or methods identified for that use. This is a special concern for directly buried conductors; they must be spliced with devices approved for that use. Frequently, conductors, lugs, and devices may have different temperature ratings. In these cases, the ampacity rating of the circuit must be consistent with the lowest temperature rating of any item in that circuit. Otherwise, one or more of the items could be operated above its safe temperature range.

- *110.26. Spaces about Electrical Equipment.* Sufficient access and working space is required around all electric equipment. This is necessary so that it can be safely serviced. Electrical equipment that is accessible only by lock and key is considered to be "*accessible to qualified persons.*" This is a fairly important statement, as "accessible only to qualified persons" shows up repetitively in the *NEC*. The table associated with this section is a very important one. You should be familiar with it.
- *110.27. Guarding of Live Parts.* Live electrical parts (*live* means energized) must be guarded against accidental contact. Again, this is a commonsense rule but one that is worth stating and defining. For example, is it necessary to guard parts that operate at 12 volts? The *NEC* says that any parts operating at 50 volts or more must be guarded, unless another section of the code specifically allows the part to be unguarded for a certain type of installation. The methods of protection are itemized in this section as well and include being in a separate, locked room, being behind some sort of partition, or being elevated above a level where casual contact is possible.

Impossibilities

From time to time the *NEC* will allow, or even require, products, constructions, or materials that may not be available at the time the *Code* is adopted. When this happens, the authority having jurisdiction is expected to permit the use of the products or techniques that comply with the previous edition of the *Code*.

This is done when the *Code* writers want to precede a product to market or to push the market in a certain direction.

Enforcement

Before you can legally occupy a house, a local government must declare it fit for habitation. Among the inspections that are required for a Certificate of Occupancy is an electrical inspection. To obtain this inspection, you must first secure (that is, purchase) a permit to install electrical wiring. Every municipality has its own process for acquiring this permit, but it generally entails going to City Hall, submitting a set of wiring plans, talking to the inspector, and paying a fee. (In most places, an individual is still allowed to wire his or her own home, but not everywhere.)

After obtaining a permit, two or three inspections are usually required. The local inspector, of course, will dictate the exact number, but two or three are fairly standard. The first inspection is usually called the *rough inspection*. This is an inspection of wiring inside the walls, before the wall surfaces are installed. Typically, this involves the inspector checking all of your in-wall wiring before the drywall goes up. The second inspection is usually called *trim-out* or *trim*. This is an inspection of the final wiring system and occurs after the walls are complete; all the switches, receptacles, and lights have been installed; and so on.

There are also inspections that may be required for under-floor circuits. This inspection occurs before concrete slabs are poured. The inspector will check any wiring to be installed and then give or deny permission for the concrete to be poured.

Another inspection is sometimes required before energizing the electrical service to the house. This frequently coincides with the rough inspection, but not always. There may be an inspection required to set up a temporary power service to the construction site, but this is usually handled separately in the construction process.

Defining Acceptable

After knowing what types of inspections are required, the next question is, "What will they require?" In almost all cases, the basic rules are those of the *NEC*. That being said, you must verify this with your local inspector, as he or she may have supplementary rules.

The *NEC* is written so as to be acceptable as law. Most municipalities pass a simple ordinance making the *NEC* law in that territory. The critical player in the *Code* being used as law is the authority having jurisdiction for enforcement of the *Code*. In most cases, the authority is the local electrical inspector. Once the *Code* is law for that specific territory, this authority will have the responsibility for making interpretations of the rules, for the approval of equipment and materials, and for granting any special permissions. The authority having jurisdiction may, if he or she desires, waive specific *Code* requirements and may permit alternate methods of installation, so long as safety is maintained.

You can see that the authority specified by the *NEC* (again, usually a local electrical inspector) has a great deal of power over electrical installations. There is no direct recourse to this person's rulings. That being said, there has been relatively little abuse of this power, at least as far as safety compromising is concerned. If an inspector were over-using his or her authority, the only recourse would be to the inspector's superior or local law-making assembly.

Chapter 2

Electrical Services

Electrical services are the first portions of a house wiring to be designed and are the most critical part of any wiring system. They must be designed and installed correctly if the overall system is to function well. All current in the house flows through the service.

In addition, services are the portion of the wiring system that are most exposed to dangers. A service is connected directly to utility company lines, which are fused at very high levels and are frequently capable of delivering thousands of amperes. That is enough capacity to cause a large explosion or fire. In addition, services are exposed to direct lightning strikes and to massive induced voltages from nearby lightning strikes. (Induced voltages are caused by the lightning strike's powerful magnetic field cutting through the service conductors.)

If services are not designed properly, installed carefully, and constructed with quality materials, there are risks.

Most inspectors will start any inspection at the service point. It is here that clues will be found as to what to expect in other portions of the wiring system. The service and service equipment, plus the grounding, are the watchdogs of the rest of the wiring system. The protection against overloads and faults is found there, as well as grounding for the protection of the system from shocks, lightning, breakdown of the transformer windings, and other damaging conditions.

Service entrances, service equipment, and the grounding of services are all extremely important subjects. In this chapter, we will cover these subjects carefully. It is critical that a service be installed properly.

Service Basics

An electrical *service* is the portion of the supply conductors that extends from the street main (or duct or line transformer) to the entrance or service panel inside the house. There are various methods of making a service entrance into a building, and they may be classified as conduit or underground services.

Following is a set of definitions for services. It is important to be clear on these before proceeding.

> Service. *The conductors and equipment for delivering energy from the electricity supply system to the wiring system of the premises served.* This definition is very complete and applies to

all wiring and equipment extending from the last pole or underground system through the service equipment. The following definitions will give the breakdown of the separate parts or sections of a service.

Service conductors. *The supply conductors that extend from the street main or from transformers to the service equipment of the premises supplied.* Therefore, service conductors are the conductors defined under Service above.

Service cable. Service conductors in the form of a cable.

Service drop. *The overhead service conductors from the last pole or other aerial support to and including the splices, if any, connecting to the service-entrance conductors at the building or other structure* In rural areas, the utility company often locates a meter pole in the yard; the meter pole may or may not have an overcurrent device installed. The service drop does not stop at the meter pole but continues on to the building or buildings or other structures that it serves (Figure 2-1).

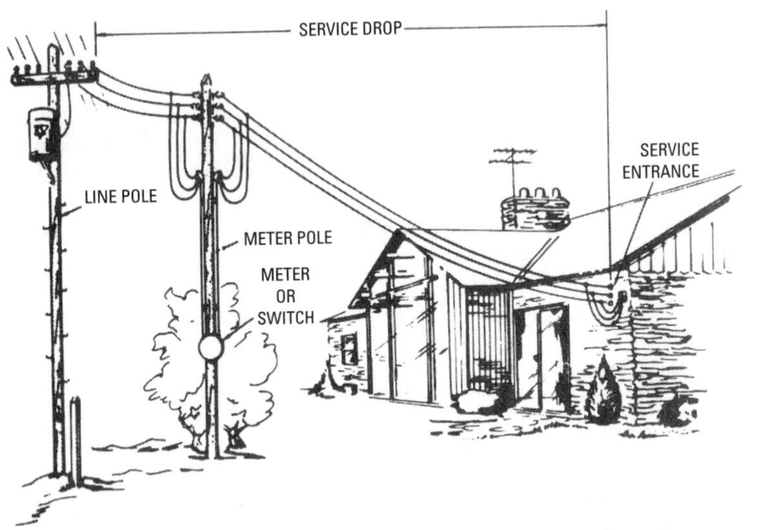

Figure 2-1 The service drop attached to a building or other structure.

Service-entrance conductors, overhead system. The service entrance includes the conductors from the service equipment to a point outside the building, clear of the building walls. The conductors are attached to the service drop at this point by either a tap or a splice. If on the building wall, the meter housing and meter are not considered as parts of the service-entrance equipment. See Figure 2-1 for a sketch showing some possible conditions.

Service-entrance conductors, underground system. The service conductors between the terminals of the service equipment and the point of connection to the service lateral, as shown in Figure 2-2.

(FPN): Where service equipment is located outside the building walls, there may be no service-entrance conductors, or they may be entirely outside the building.

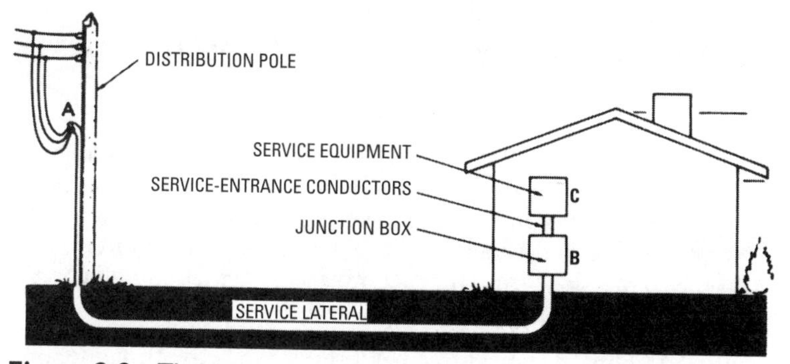

Figure 2-2 The service lateral extends from point A to point B. The service entrance is from point B to point C.

Service equipment. Necessary equipment usually consists of a circuit breaker or fuses and a switch located on the inside or outside of the building near the point of entrance. The service equipment is intended to constitute the means of disconnecting the electrical supply entering the building, as shown in Figure 2-3. Further information is covered in *Article 230.*

Service lateral. The service lateral includes the underground service conductors, including any risers up the pole at the street main or transformer structure. The conductors are considered as service laterals until they enter a junction box in the building. If such a box is not used, they will cease to be service laterals at the point of entrance into the building, at which they become service-entrance conductors. If the service-entrance equipment is located on the outside of the building, there may be no service-entrance conductors; the conductors could all be termed service laterals (Figure 2-2).

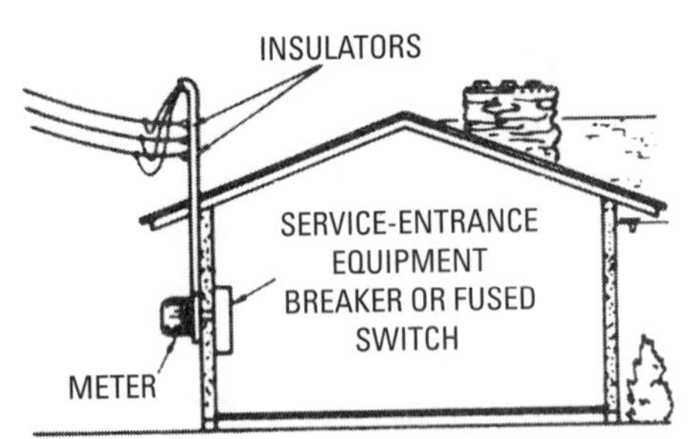

Figure 2-3 Service-entrance equipment that serves as the electrical supply disconnect.

Service raceway. This is any raceway, conduit, or tubing enclosing the service-entrance conductors. Where a service mast is used, the conduit to the metering circuit, the raceway (for connecting to the metering if several should be required), and the connections from the raceway to the service equipment are all considered the service raceway.

Figure 2-1 shows a typical overhead service drop, with Figure 2-4 illustrating the minimum clearances over certain areas, as required by the *NEC*.

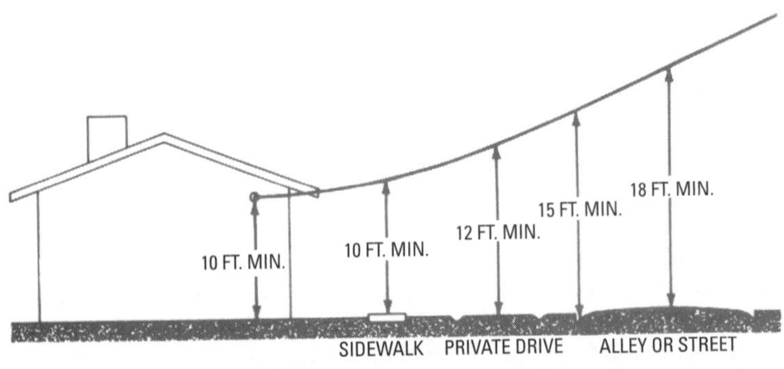

Figure 2-4 The *NEC* calls for minimum clearances above certain aspects of the landscape. Note here that the service must be no less than 10 feet above ground level where it enters the residence.

Overhead Service Components

Many utility companies now install meters outside residences for easier reading. Where a meter pole is not used, the meter is set on the outside of the house. The service wires are then connected through more conduit that passes through the wall and through a bushing to the *entrance panel*, from which power is distributed to the house through the various circuits (Figure 2-5).

With reference to Figure 2-6, rigid conduit is used from the cutout switch to a point at least eight feet above the ground. The wires enter the conduit through a fitting called a *service cap* in order to protect the wires at the entrance point and prevent water from entering the conduit. (Note that the meter in Figure 2-6 is installed inside the home.) A Pierce wire-holder insulator may be used, as shown in Figure 2-7.

To install a conduit service entrance, a hole is drilled through the wall to pass the conduit. The conduit is then bent so that the end passing through the wall extends 3/8 inch inside the main switch cabinet. Instead of bending the conduit, an approved L (condulet) fitting is often used, as shown in Figure 2-8.

The end of the conduit is secured to the entrance panel box by a locknut and bushing. The locknut is screwed onto the conduit before it enters the panel. The bushing protects the wires where they leave the pipe, and it should be tightened with a pair of pliers. The locknut is then tightened against the wall of the cabinet to hold the conduit securely in the box.

That portion of the conduit that is on the outside of the building is held in place by pipe straps, which in turn are fastened with screws. The L condulet must be of the weatherproof type. Figure 2-9 shows one type of L condulet. This fitting is made weatherproof by placing a rubber gasket between the body of the fitting and the cover.

Residential Underground Service

There is a considerable trend to underground service in densely populated areas. An underground service lateral (wire or cable extending in a horizontal direction) is installed, owned, and maintained by the public utility. It is run to the residential termination facility, which is usually the meter or meter enclosure, as depicted in Figure 2-10. Meters are ordinarily located within 36 inches of wall nearest to the street or easement where the public utility's distribution facilities are located. The meter is mounted from 48 to 75 inches above final grade level. If there is likelihood of damage,

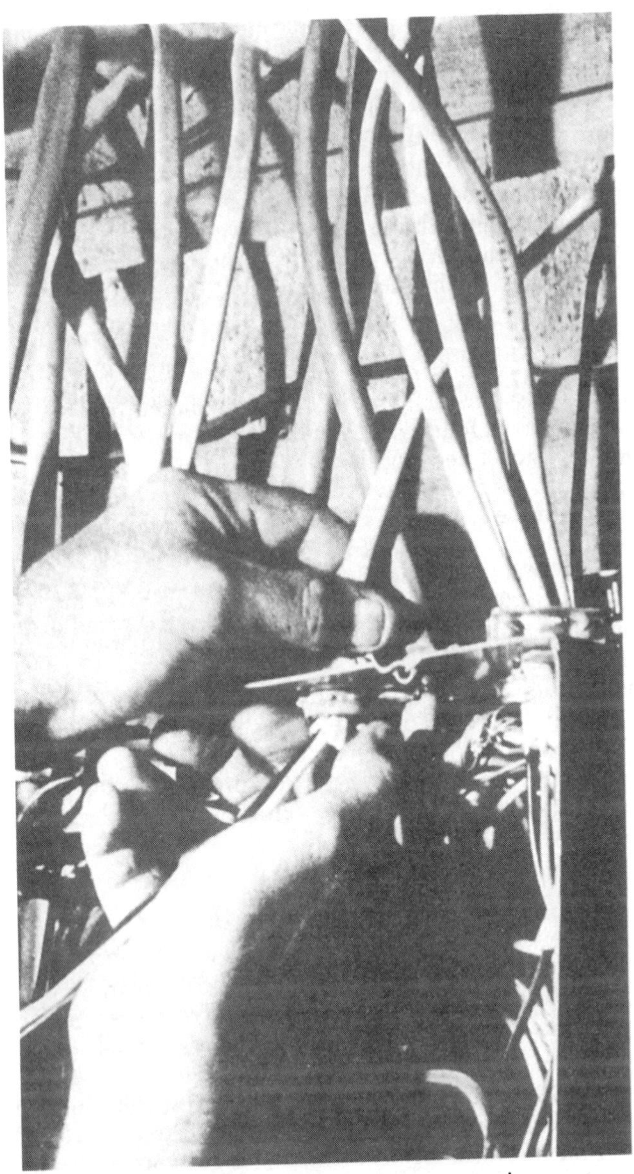

Figure 2-5 Service entrance panel.

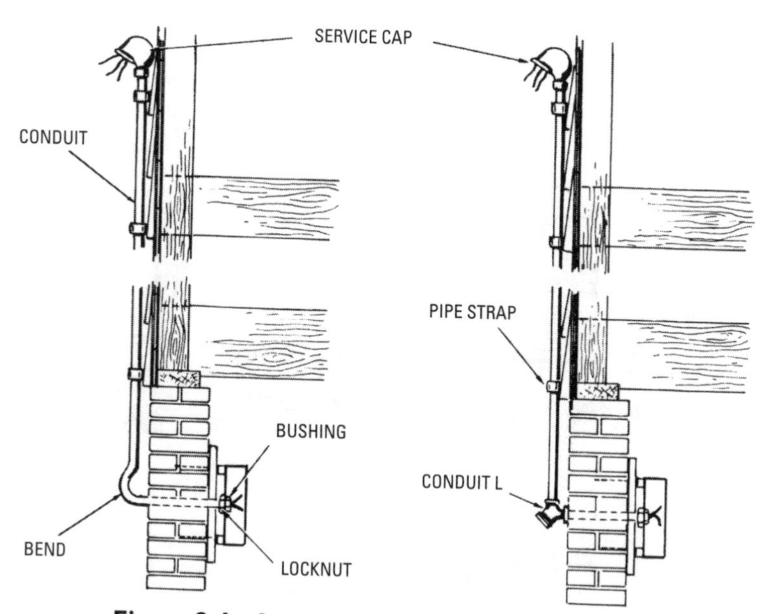

Figure 2-6 Service entrance cap and installation.

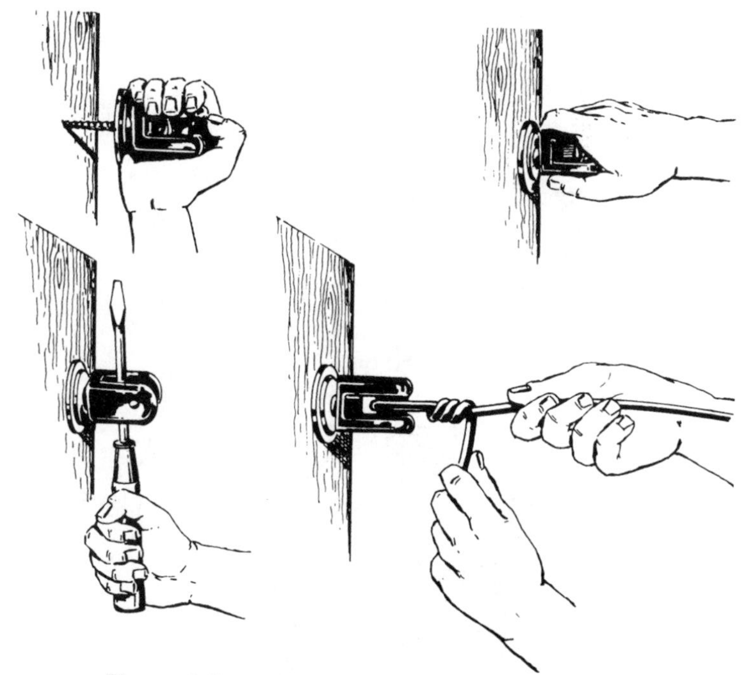

Figure 2-7 Installing a Pierce wire-holder insulator.

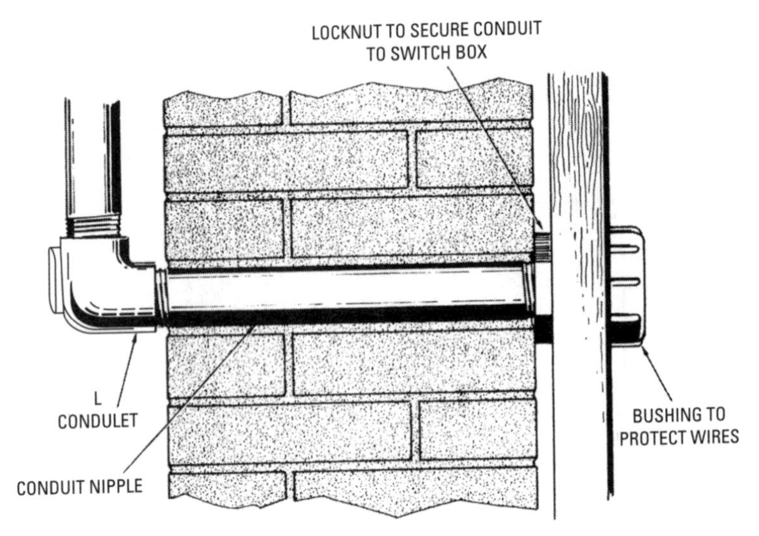

Figure 2-8 Service run with L (condulet) fitting.

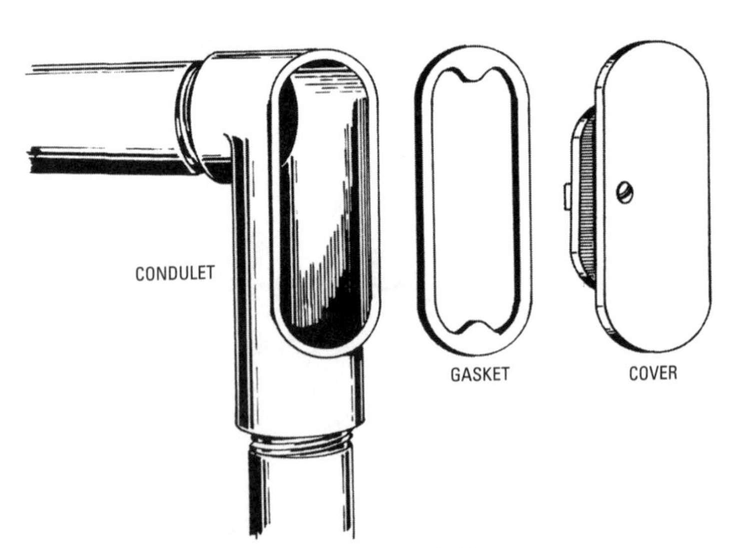

Figure 2-9 Weatherproof mogul condulet.

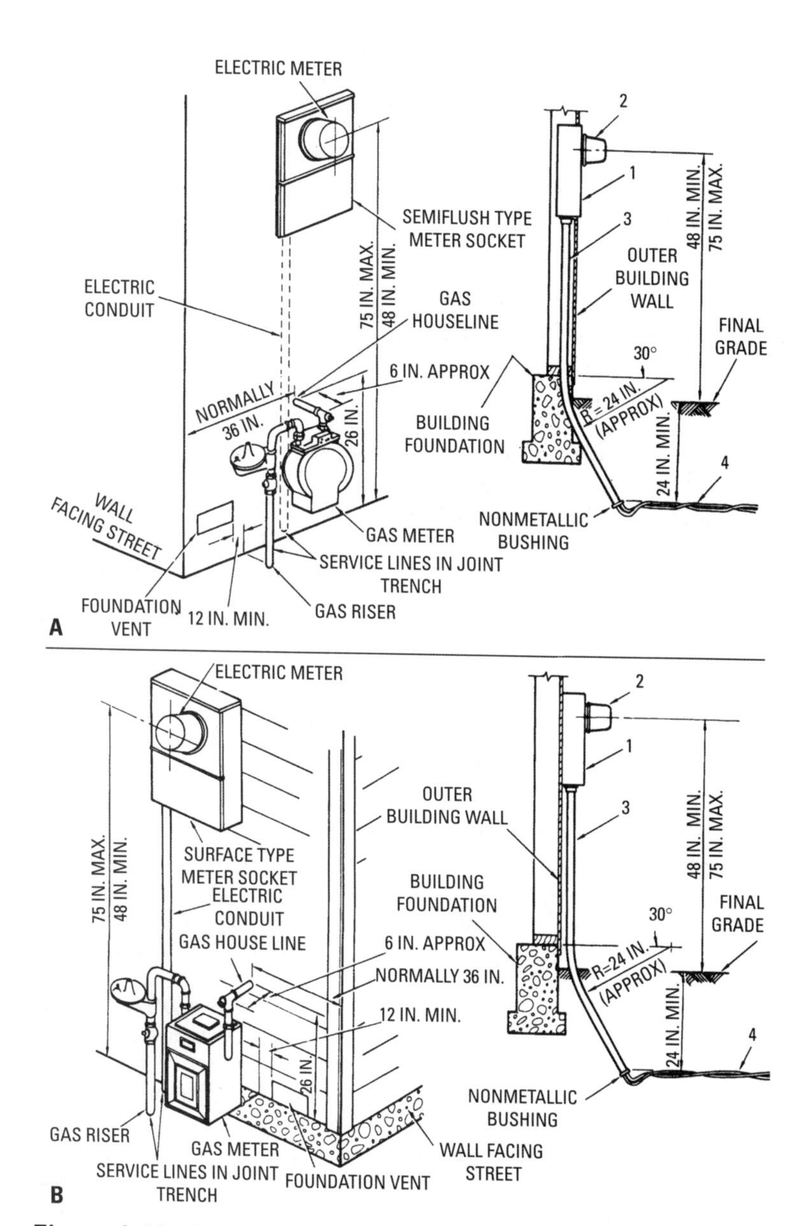

Figure 2-10 Underground residential service-entrance installations: (A) recess-mounted service termination enclosure and (B) surface-mounted termination enclosure.

meters must be adequately protected. Larger conduit is now required than formerly, and public utilities are frequently specifying 2-inch minimum inside diameter conduit for the service entrance. Note that aluminum conduit cannot be installed below ground level. The service conductors are usually No. 2 aluminum wire, and the service is customarily three-wire 120/240-volt single-phase 60 Hz. The third wire is a grounded neutral conductor.

NEC Requirements for Services

Almost all of the requirements that pertain to electrical services are contained in *Article 230* of the *NEC*. However, many of the requirements of *Article 230* apply to industrial establishments, commercial structures, or other applications. Here, we cover only the requirements that apply to house wiring.

Concerning the minimum size of service drop conductors, *Section 230.23* of the *NEC*, *Size and Rating of Service Drop Conductors*, states: *Conductors shall have sufficient ampacity to carry the load. They shall have adequate mechanical strength and shall not be smaller then No. 8 copper or No. 6 aluminum or copper-clad aluminum.* An exception is included but is not pertinent to this discussion.

Concerning underground service conductors, *Section 230.31* of the *NEC* states: *Conductors shall have sufficient ampacity to carry the load. They shall not be smaller than No. 8 copper or No. 6 aluminum or copper-clad aluminum.* Again, an exception is listed that is not pertinent to this discussion.

Section 230.79(C) states that for single-family residences with an initial load of 10 kW or more computed in accordance with *Article 220*, or if the initial installation has six or more two-wire branch circuits, the service-entrance conductors shall have an ampacity of not less than 100 amperes for three-wire service.

These sections, especially *Section 230.79*, tell us a great deal. In broad terms they indicate that 100-ampere service is the minimum permitted for a single-family residence. Note the word minimum. Larger services are often required, as in large residences and in electrically heated residences. Also remember that when considering the *NEC*, minimum requirements are always discussed. Wiring is not installed just for today, but for anticipated future needs. Be sure that the service size meets the requirements of *Article 220* with a 100-ampere ampacity (or larger if the calculations require a larger service).

Table 2-1 Conductor Types and Sizes—RH-RHH-RHW-THHW-THW-THWN-THHN-XHHW

Copper	*Aluminum and Copper-Clad Aluminum*	*Service Rating in Amperes*
AWG	AWG	
4	2	100
3	1	110
2	1/0	125
1	2/0	150
1/0	3/0	175
2/0	4/0	200

For single copper-conductor service drops in free air, use *Table 310.17*. For aluminum conductors, use *Table 310.19*. Note that the ampacities in these two tables are greater than those in *Tables 310.16* and *310.18*. This is because of the greater heat dissipation in free air.

Note 3, following and pertaining to *Tables 310.16* through *310.19*, gives alternate ratings for service-entrance conductors:

> Three-Wire, Single-Phase Dwelling Service. *In dwelling units, conductors, as listed* [in *Table 2-1* in this chapter], *shall be permitted to be utilized as three-wire, single-phase, service-entrance conductors and the three-wire, single-phase feeder that carries the total current supplied by that service.*

Service-Entrance Location

Where should the service entrance and service equipment be located? This is sometimes quite an involved problem, but not in all cases.

First of all, the serving utility is concerned with where in the residence a service drop is to be installed. In order to supply power to the house efficiently, the service drop may need to be on one particular side of the home. Service laterals do not seem to cause as much of a problem, because they do not have to run in a direct line.

Consider all of these items when locating the service entrance and the service equipment:

According to *Sections 230.70(A)* of the *NEC, the disconnecting means shall be located at a readily accessible point nearest to the entrance of the conductors, either inside or outside the building or structure. Sufficient access and working space shall be provided about the disconnecting means.*

Section 230.6 (Conductors Considered Outside Building) states that *conductors placed under at least 2 inches of concrete beneath a building, or conductors within a building in conduit or duct and enclosed by concrete or brick not less than 2 inches thick, shall be considered outside the building.*

Inspection authorities differ on the length of the service-entrance conductors to the service equipment. Check with the local authority having jurisdiction. Refer to *Section 230.70(A)*. This section states a great deal in very few words. What does the phrase "*readily accessible*" mean? To the author, it indicates that the disconnecting means shall not be in a bedroom, bathroom, dish cupboard, or other enclosed area, and a disconnect in the basement will not be accepted unless there is a ground-level exit. The disconnecting means is an emergency item, and the author is certain that a basement would be the last place he would wish to go in the event of a fire. Bedrooms and bathrooms are private rooms, and cupboards obviously have restricted access. A bathroom has grounded items, in addition to the steam and moisture present, which could create a hazard.

Take note of the phrase "*nearest the point of entrance of the service conductors.*" Here the inspector must use his powers of interpretation as granted in *Section 90.4*. To the author this means not to exceed a distance of approximately 15 feet. Even 15 feet might be too far in some cases, and would require overcurrent protection at the outer end. If in doubt, check with the authority having jurisdiction.

Sufficient working space shall be provided in the vicinity of the service disconnecting means. Basically, this means that clearance must be provided so that an electrician does not have to lean across a washer, dryer, or other appliance in working on the equipment, or have to use a ladder or chair, or have any obstruction in the way when working on the panel.

Section 230.90 states that the service-entrance conductors shall be the same ampacity as the main or larger.

Installation of Service Drops

If a service drop is used, check the point of attachment to make sure that all requirements of *Section 230.24* in the *NEC* are met.

> *230.24. Clearances. Service-drop conductors shall not be readily accessible and shall comply with (A) through (D) below*

for services not over 600 volts, nominal. [only section (*A*) is reproduced here]

(A) Above Roofs. Conductors shall have a vertical clearance of not more than 8 feet (2.44 m) from the roof surface. The vertical clearance shall be maintained for a distance of not less than 3 feet (914 mm) in all directions from the edge of the roof. [See Figure 2-11.]

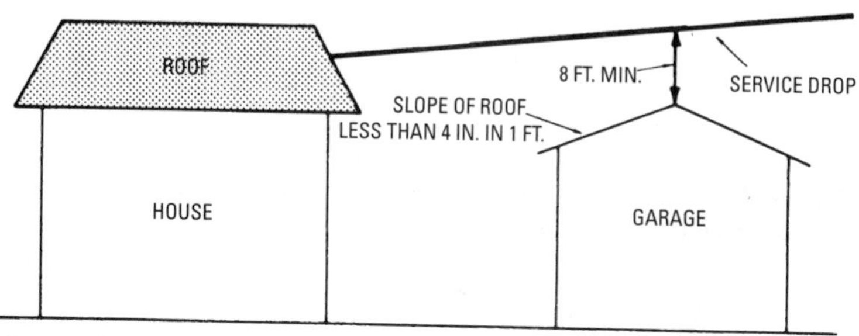

Figure 2-11 Clearance of conductors passing over roofs must conform to Code rulings.

Exception No. 1: The area above a roof surface subject to pedestrian or vehicular traffic shall have a vertical clearance from the roof surface in accordance with the clearance requirements of Section 230.24(B).

Exception No. 2: Where the voltage between conductors does not exceed 300 and the roof has a slope of not less than 4 inches (102 mm) in 12 inches (305 mm), a reduction in clearance to 3 feet (914 mm) shall be permitted. [See Figure 2-12.]

Exception No. 3: Where the voltage between conductors does not exceed 300, a reduction in clearance above only the overhanging portion of the roof to not less than 18 inches (457 mm) shall be permitted if (1) not more than 4 feet (1.22 m) of service drop conductors pass above the roof overhang, and (2) they are terminated at a through-the-roof raceway or approved support [Figure 2-13]. *See Section 230.28 for mast supports.*

Exception No. 4: The requirement for maintaining the vertical clearance 3 feet from the edge of the roof shall not apply to the final conductor span where the service drop is attached to the side of a building.

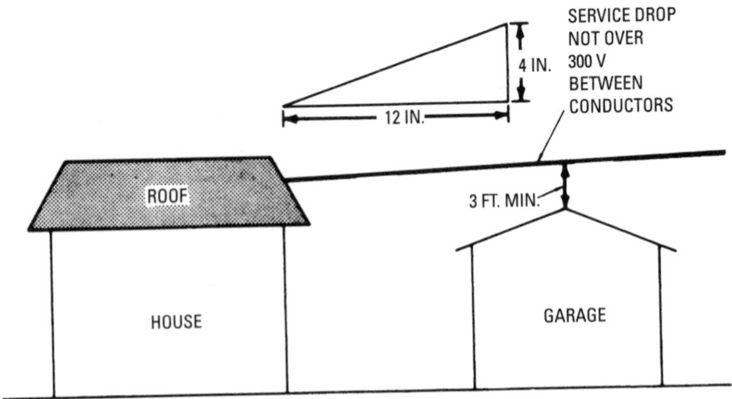

Figure 2-12 Clearance of conductors passing over roofs is governed by voltage and by the roof slope.

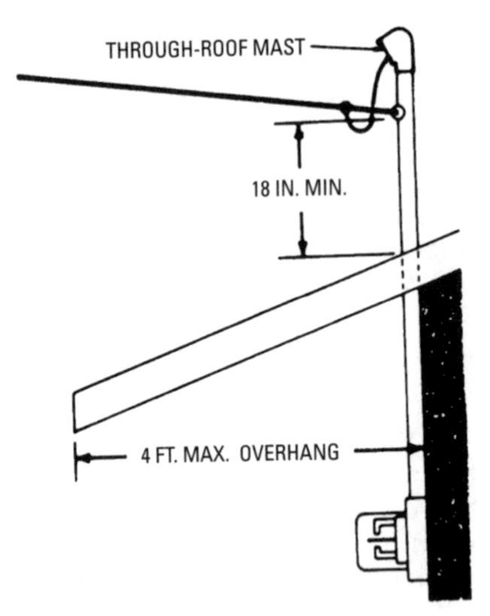

Figure 2-13 Conductors passing over a portion of a roof and terminating at a through-the-roof service raceway have different clearance requirements.

Vertical clearance from the ground is covered in *Section 230-24(B)* of the *NEC*. The section will not be reproduced here but is illustrated by Figure 2-14.

Clearance from building openings is covered by *Section 230.9* of the *NEC*. The intent is shown in Figure 2-15.

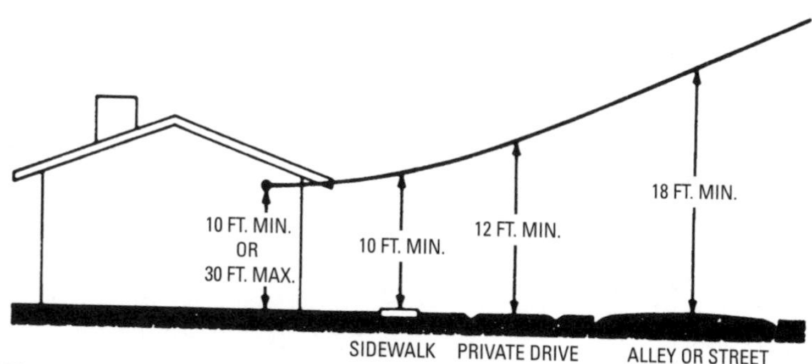

Figure 2-14 Minimum service-drop clearance.

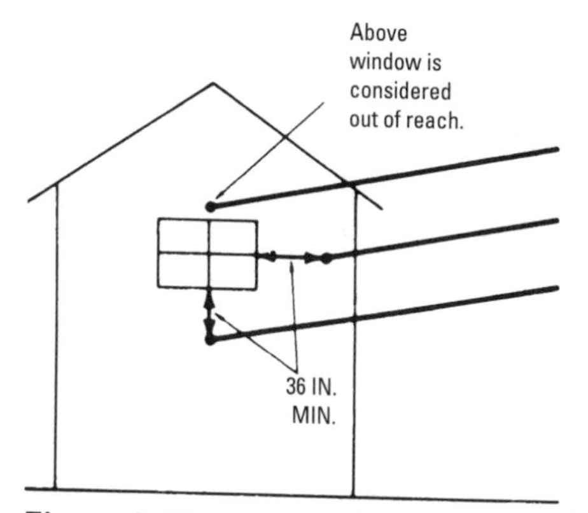

Figure 2-15 Minimum service-drop clearance around building openings.

If a mast is required for clearances, refer to *Section 230.28* of the *NEC*. This mast shall be strong enough to support the service drop in sleet storms, high winds, and other stress conditions, or it will require guying. In the author's inspection area, nothing smaller than 2-inch rigid galvanized conduit is accepted, and guying is required according to the judgment of the inspector (Figure 2-16).

For clearance from swimming pools, *Section 230.24(D)* refers you to *Section 680.8*.

Installation of Service Laterals

Service laterals may be direct-burial conductors or cable or may be in approved raceways. Refer to *Section 230.30*.

> *230.30. Insulation. Service lateral conductors shall be insulated for the applied voltage.*

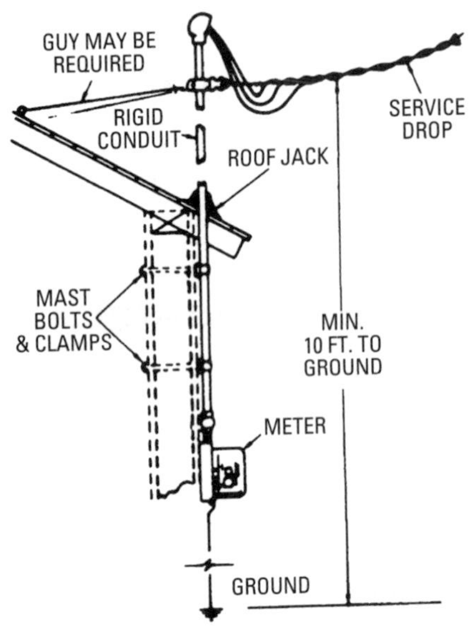

Figure 2-16 Mast installation for proper service-drop height.

Exception: A grounded conductor shall be permitted to be uninsulated as follows:

A. Bare copper used in a raceway.

B. Bare copper for direct burial where bare copper is judged to be suitable for the soil conditions.

C. Bare copper for direct burial without regard to soil conditions when part of a cable assembly identified for underground use.

D. Aluminum or copper-clad aluminum without individual insulation or covering when part of a cable assembly identified for underground use in a raceway or for direct burial.

Take note of the bare copper for direct burial, but also take note of the fact that the soil conditions must be judged suitable, and this judging is up to the authority having jurisdiction, as provided in *Section 90.4*.

An approved raceway may be used. This could be duct, nonmetallic conduit, or rigid galvanized conduit (provided the galvanizing is approved by the authority having jurisdiction). Most authorities will require additional corrosion protection over the galvanizing, as outlined, for example, in *Sections 344.10* and *344.14*. Underwriters' Laboratories have conducted tests on the corrosion of galvanized conduits and have established standards based on the ohms-per-centimeter resistivity. Since this test requires special testing equipment, most authorities will insist on additional corrosion protection. The service laterals may be installed in these approved raceways, but the insulation of the conductors shall meet the requirements of *Section 310.8(C), Wet Locations*. Types MTW, RHW, TW, THW, THWN, XHHW, moisture-impervious metal-sheathed (traditionally lead-covered), and other approved types, may be installed. For direct burial, the conductors shall be Type USE buried a minimum of 24 inches deep. The authority enforcing the code may require supplemental mechanical protection, such as a covering board, concrete pad, or raceway. In rocky soil, and more especially where frost is prevalent, the inspection authority will usually require a fine sand bed with a fine sand covering under and over the conductors. Rocks subjected to frost heave will cause damage to the insulation.

Mechanical protection is also required where a conductor is entering the building or leaving the ground to go up a pole. See Figure 2-17 and *Section 230.50.*

Raceway sealing is required at the building to prevent the entrance of moisture and gases. Duct seal may be used. Service-lateral

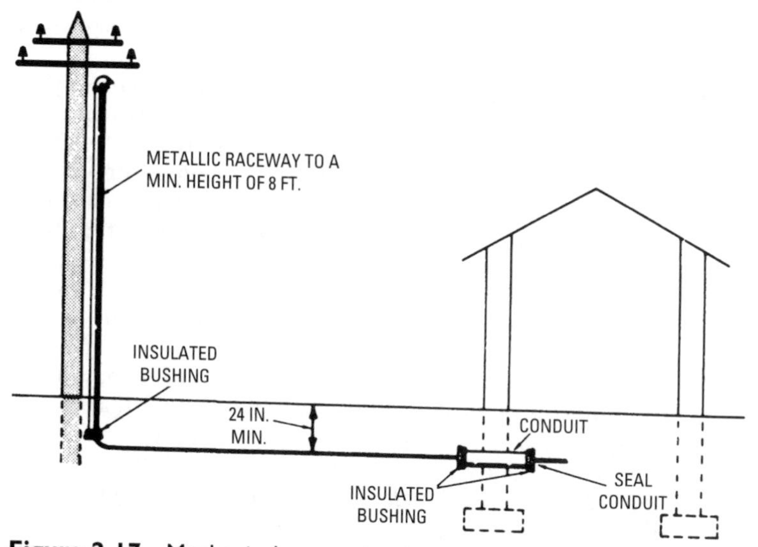

Figure 2-17 Mechanical protection is required for underground service conductors where they enter a building or leave the ground to go up a pole.

conductors shall be without splice, but where they enter the building they cease to be service-lateral conductors and become service-entrance conductors, so a splice is permitted at that point (Figure 2-18).

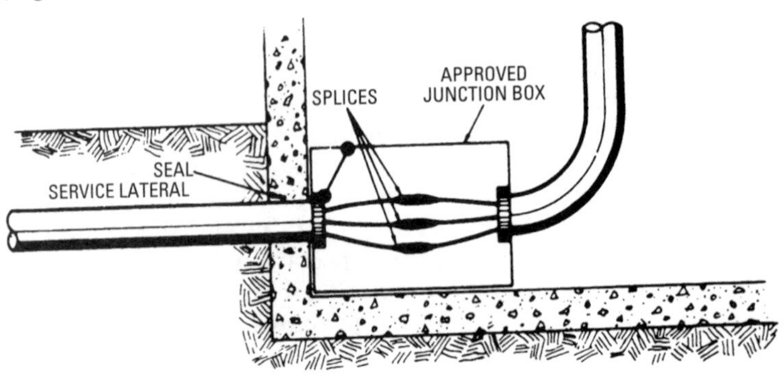

Figure 2-18 A splice in an underground system.

Service Entrances and Equipment

A portion of this subject was covered earlier in the chapter. Figures 2-13 and 2-16 showed, respectively, service masts and the conductors from the tap to the service drop, on down to the meter. From this point to the service equipment the conductors are service-entrance conductors.

Section 230.46 allows service-entrance conductors to be spliced, but except for very specific applications, this is almost never allowed. For example, these conductors will necessarily be broken in the meter box, so the *Code* permits clamped or bolted connections in meter housings. But other splicing of service conductors is, first of all, dangerous. These conductors are protected only by the utility company's fusing, which may be many hundreds of amperes—much more than the conductors themselves could withstand. Also, a short would tap power from the service ahead of the meter. This would certainly not be acceptable to any utility company.

Figure 2-19 shows service entrances other than a mast type. Where the proper height can be obtained without a mast, as shown in Figure 2-14, such an installation may be used. This might be service-entrance cable (*Article 338*), rigid conduit (*Article 344*), or electrical metallic tubing (*Article 358*), provided watertight fittings are used.

The service-entrance conductors extend from point A in Figure 2-19 to the line side of the main disconnecting means in the service

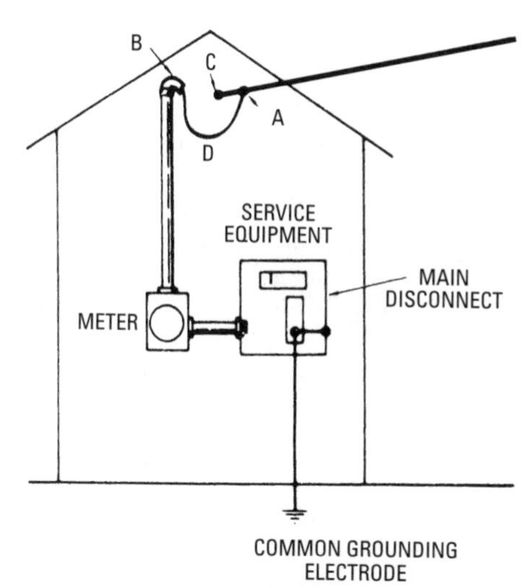

Figure 2-19 A typical service installation.

equipment. Notice that the service-entrance head (B) is higher than the point of attachment of the service drop. This is covered in *Section 230.54., Service Cable Equipped with Raintight Service Head or Gooseneck*. If it is not possible to install the service head higher than the point of attachment, there are other alternatives. However, the basis of all such practices is to keep water from running into the service raceway and equipment. Notice the drip loop (Figure 2-19, part D) for this purpose.

Some utilities will require more than the minimum 10 feet height for the point of attachment allowed by the *Code*. In such cases, a minimum of 12 feet for the point of attachment is generally used.

How high should the meter housing be installed? This depends upon the utility requirements, but in most cases it will not be less than 5 feet or more than 6 feet high.

Basically, service equipment is defined as the circuit breakers, fuses, or switches and their accessories to be used as a main disconnect to the residence. *Section 230.70(A)* states that the service equipment shall be on the inside or outside of the building or structure "*nearest point of entrance of the service conductors.*" In the majority of cases, the main disconnect for a residence will be in an enclosure along with the branch circuit breakers or fuses. A raintight main disconnect may be installed on the outside of the house to serve as the main disconnect, with a feeder circuit from this main to a branch-circuit panel in the house. If this is the case, the conductors from the main to the branch-circuit panel are feeders and shall have an ampacity equal to or larger than the main disconnect. Also, an equipment ground conductor from the main to the branch circuit is required, with the neutral insulated from the enclosure of the branch-circuit panel. This equipment-grounding conductor may be a metal raceway (conduit or EMT) or, if the feeder is a cable, this cable shall have an insulated neutral, two-phase conductors, and an equipment-grounding conductor (Figure 2-20).

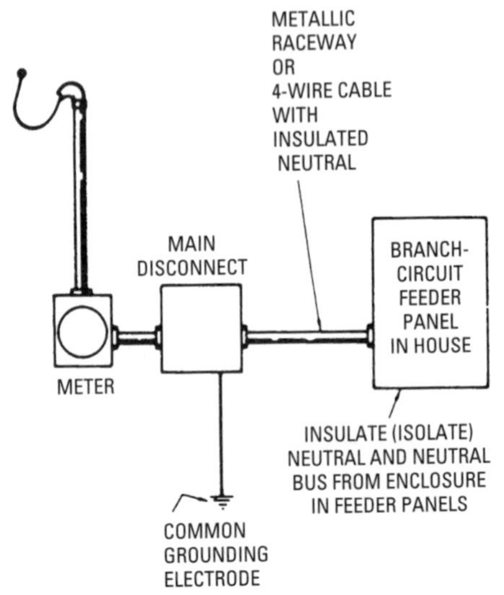

Figure 2-20 Proper installation of a raintight disconnect on the outside of a house.

The branch-circuit panel fuses shall have an ampacity equal to or greater than the rating of the main disconnect breaker, switch, or fuses. If the branch-circuit fuses of a feeder panel are not of this capacity, then a main breaker shall be installed to protect the fuses. The only place you might run into this in a residence is where there are two or more feeder panels. This would only be the case in a very large residence. There is nothing in the *Code* prohibiting the installation of a branch-circuit panel with a main disconnect on the outside of the residence, provided a raintight enclosure is used. See Figures 2-20 and 2-21.

Recall that if the initial installation has six or more two-wire branch circuits, the service-entrance conductors shall have an ampacity of not less than 100 amperes for a three-wire service.

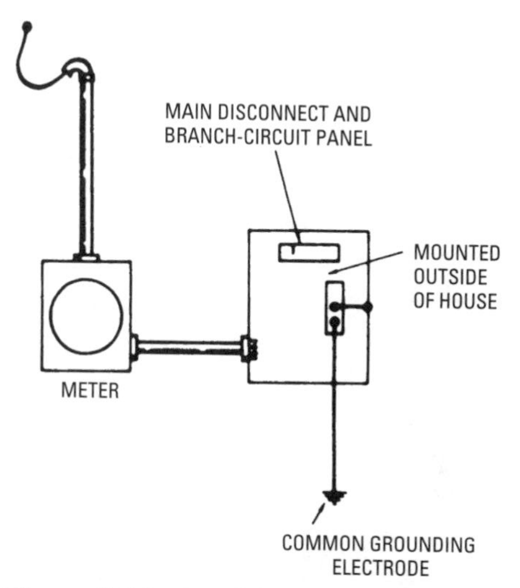

Figure 2-21 Installation of a branch-circuit panel and main disconnect on the outside of a house.

You must post an accurate circuit directory at the service panel. All panels are sold with a blank directory, but you must fill it out correctly and in detail. Identify each circuit and the devices it feeds.

It should be mentioned here that white-colored conductors shall never be used for phase conductors—neither may white-colored conductors be marked with some other color. White conductors are to be used strictly as the grounded conductor (neutral). This applies to a gray color as well. However, *Section 200.6* permits insulated conductors larger than No. 6, other than white or gray, to be used as the grounded conductor (neutral), provided they are plainly identified at the terminations with white at the time of installation.

There are some utilities that require one meter for the residence proper and one for the water heater, with an off-peak time clock and a different rate structure. When this is required by the utility, you will no doubt find that they will require the same type of installation as does the *Code* (Figure 2-22).

Grounding of Services

Grounding is covered by *Article 250* of the *NEC* and, in covering this portion of services, we should become familiar with the terminology and definitions pertaining to grounds. See the definitions in *Article 100.*

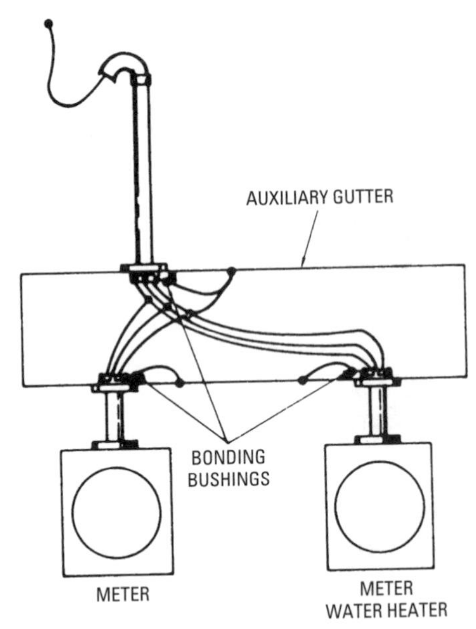

Figure 2-22 Some utilities require a separate meter for a water heater.

Ground: A conducting connection, whether intentional or accidental, between an electrical circuit or equipment and the earth, or to some conducting body which serves in place of the earth.

Grounded Conductor: A system or circuit conductor that is intentionally grounded. This would be what we term the neutral. Residences are wired with 120/240 volts or 208Y/120 volts (Figure 2-23).

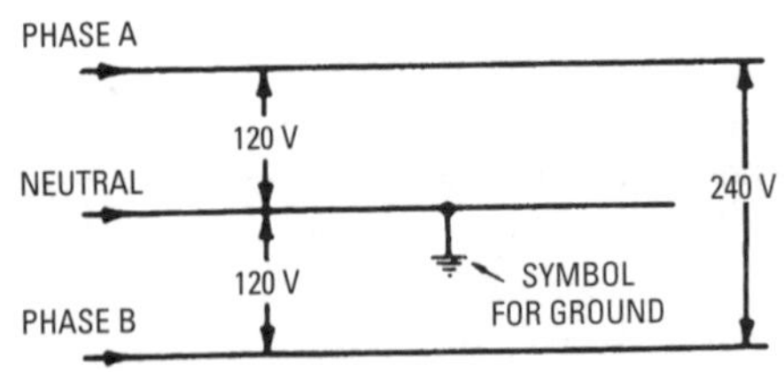

Figure 2-23 120/240-volt, single-phase, three-wire system.

Grounding Conductor: A conductor used to connect equipment or the grounded circuit of a wiring system to a grounding electrode or electrodes.

Grounding Electrode Conductor: The conductor used to connect the grounding electrode to the equipment-grounding conductor and/or to the grounded conductor of the circuit at the service equipment or at the source of a separately derived system.

Grounding of services, as previously stated, is very important and must be done properly. Different situations will be covered to give examples of various problems that might be encountered.

The basic requirements for grounding are that the path to ground from circuits, equipment, and conductor enclosures shall: (1) be permanent and continuous; (2) have capacity to conduct safely any fault current likely to be imposed on it; and (3) have sufficiently low impedance to limit the voltage to ground and to facilitate the operation of the circuit protective devices in the circuit. Part (3) is often ignored, probably because of the word impedance. *Impedance* is AC resistance. *Section 110.10* in the *NEC* emphasizes its importance. Fault currents available have generally increased in amplitude over time. This refers to phase-to-phase faults, phase-to-ground faults, and to both bolted and arcing faults. Part (3) suggests being ever mindful of impedances in connection with grounding, and to keep these impedances as low as possible.

The grounding conductor (common main grounding connector) of a wiring system is also used to ground equipment, conduit, and the supply side of the disconnecting means. This is done so that when the disconnecting means is opened, the grounded conductor will not be opened and interrupt the grounding on the system. On a service of high capacity, it is recommended that the grounding conductor be connected within the service-entrance equipment. It should always be connected within the service equipment. If the grounding connection were within the meter housing, it would not be available to an electrician doing service work (Figure 2-24).

There have been occasions when the telephone grounding conductor or an antenna ground was run into the service equipment. This should never be permitted. It is proper to tie them to the same common grounding electrode, but they should never be in the same enclosure (Figure 2-25).

Grounding Electrode

This portion of the *Code* is the basis and requirements for a properly installed grounding system. Great care should be taken to understand the contents of this portion of the book thoroughly and to adhere to the installation of the grounding system as required by the *NEC.*

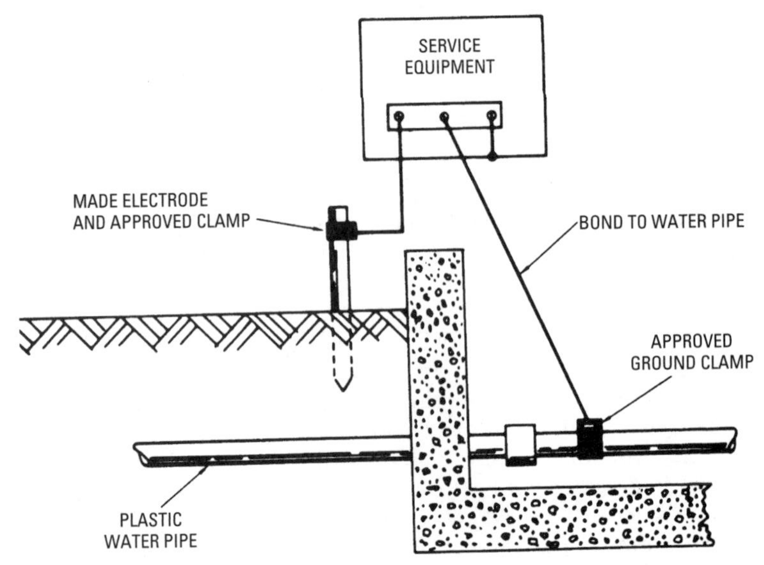

Figure 2-24 Approved method of grounding a wiring system.

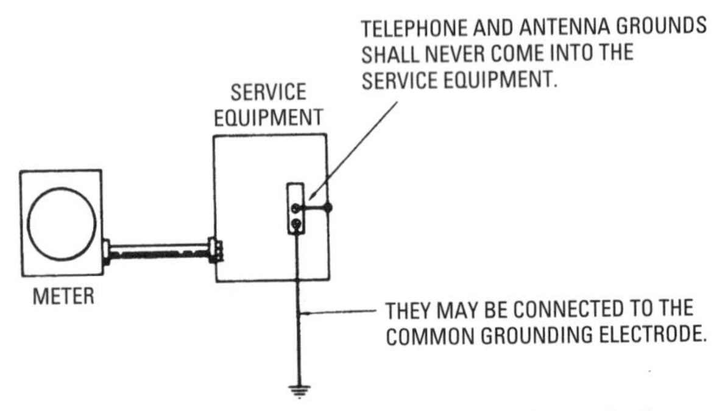

Figure 2-25 Proper method of grounding telephones and antennas.

If the common grounding conductor is connected into the meter housing, or, for that matter, into the service equipment, there shall not be an aluminum service-entrance conductor and a common grounding conductor under the same lug. The restrictions put on the use of aluminum for a common grounding conductor rule it out, so copper must be used. If the copper conductor and the aluminum conductor are in the same connector, the copper will cause electrolysis on the aluminum conductor and make a high-resistance connection (Figure 2-26).

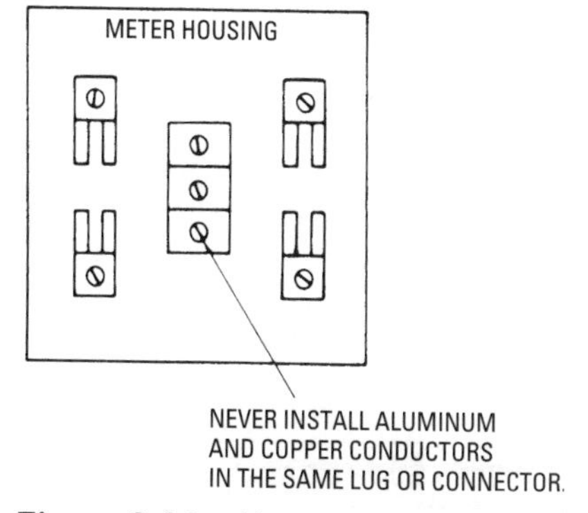

Figure 2-26 Aluminum and copper conductors must not be fastened together with the same connector.

Grounding Electrode System

This part of *Article 250* (*Section III*) is the basis for the proper construction of a grounded system. Great care should be taken to understand the contents of this part thoroughly and to adhere to the installation as required by the *Code*.

> *250.52. Grounding Electrodes.*
>
> *(A)(1) Metal Underground Water Pipe. A buried metallic underground water supply system shall always be used as the grounding electrode wherever there are 10 feet or more of buried pipe, including any well casing that is bonded to the system. If there is a chance that the piping system will be disconnected, or that an insulated coupling is or will be installed, or the possibility that nonmetallic water piping might be installed at a later date, the pipe electrode shall be supplemented by one or more made electrodes bonded to the piping.* [See Figures 2-27, 2-28, and 2-29.]
>
> *A metal underground water pipe shall be supplemented by an additional electrode of a type specified in Section 250.52.* [See Figure 2-29.] The electrical wiring system should be adequately grounded without depending on the outside piping system. This means that supplementing the water piping system with

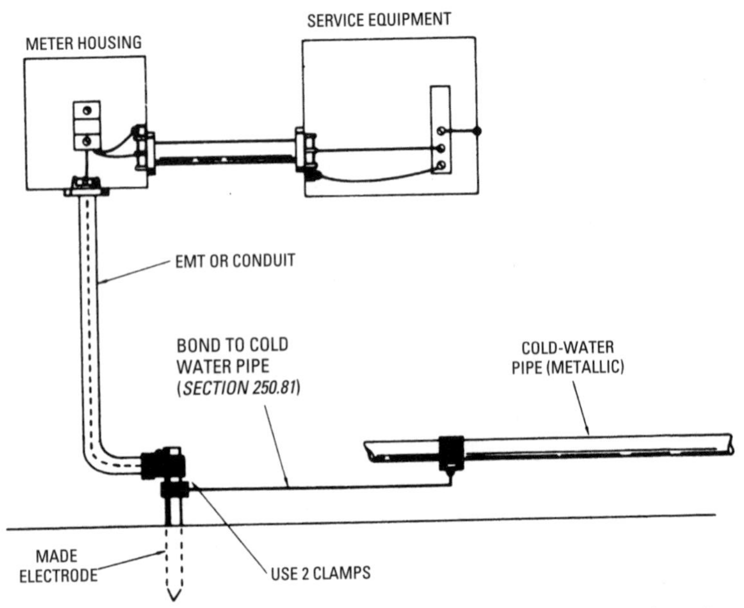

Figure 2-27 Interior cold-water piping bonded to the grounding electrode.

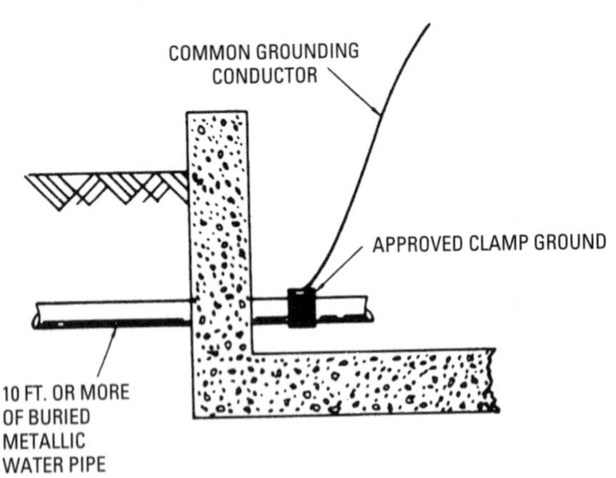

Figure 2-28 Buried metallic water pipe used as a grounding electrode.

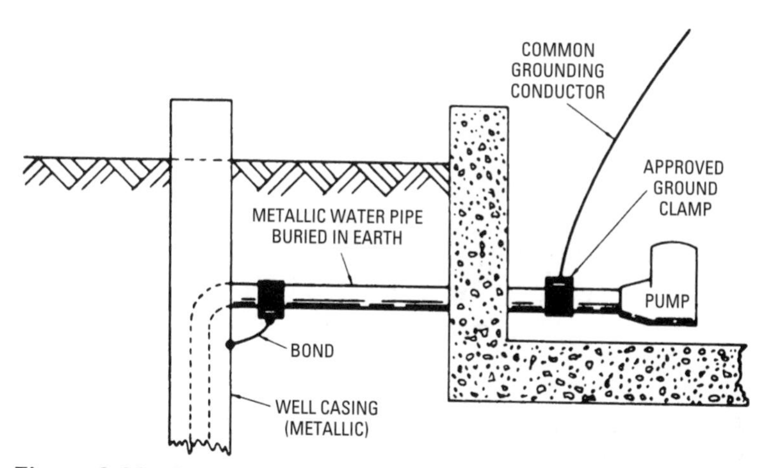

Figure 2-29 Buried metallic water pipe bonded to a well casing and used as a grounding electrode.

made electrodes is advised. Additional safety may be gained by bonding the grounding electrode to the gas, sewer, and hot-water piping and to metallic air ducts within the building.

In addition to the underground metallic pipe system, or if the metallic underground water pipe system is not available, the following may be used:

Metal Frame of the Building. The metal frame of the building, where effectively grounded.

Concrete-Encased Electrodes. An electrode encased by at least 2 inches (50.8 mm) of concrete, located within and near the bottom of a concrete foundation or footing that is in direct contact with the earth, consisting of at least 20 feet (6.1 m) of one or more steel reinforcing bars or rods of not less than 1/2 inch (12.7 mm) in diameter, or consisting of at least 20 feet (6.1 mm) of bare copper conductor not smaller than No. 4 AWG. [See Figures 2-30 and 2-31.]

Ground Ring. A ground ring encircling the building or structure, in direct contact with the earth at a depth below earth surface not less than 2 1/2 feet (7.62 mm), consisting of at least 20 feet (6.1 m) of bare copper conductor not smaller than No. 2 AWG.

Made and Other Electrodes. Where none of the electrodes specified in Section 250.52 is available, one or more of the

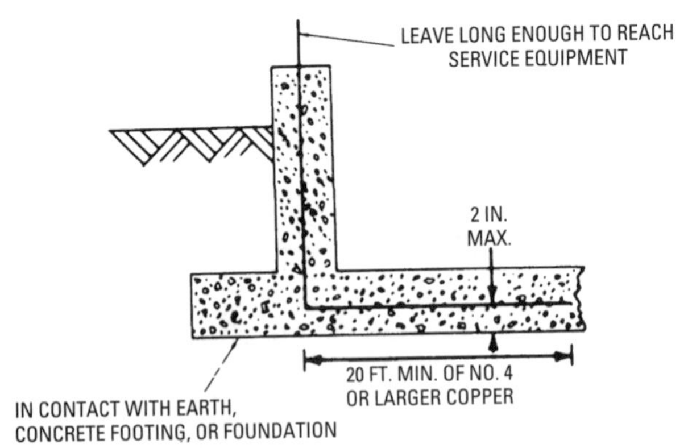

Figure 2-30 Bare copper wire encased in concrete and used as a grounding electrode.

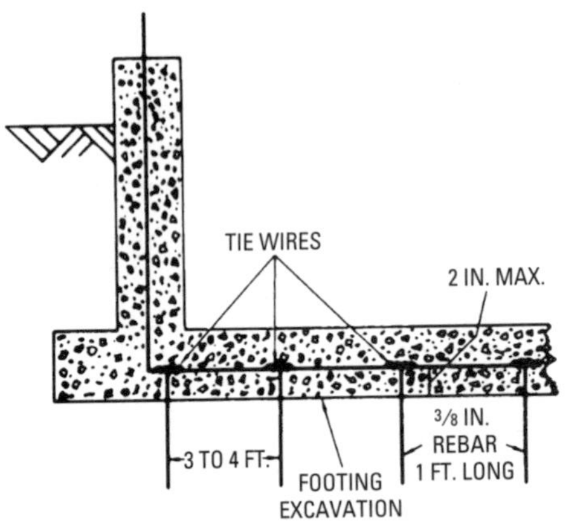

Figure 2-31 Using rebars to improve the grounding effectiveness of a copper wire encased in a concrete footing or foundation.

electrodes specified in (A) through (D) below shall be used. [(*A*) through (*D*) not reproduced here; please refer to the *NEC*] *Where practicable, made electrodes shall be embedded below permanent moisture level. Made electrodes shall be free from nonconductive coatings, such as paint or enamel. Where more than one electrode system is used (including those used for lightning rods), each electrode of one system shall not be less than 6 feet (1.83 m) from any other electrode of another system.*

Two or more electrodes that are effectively bonded together are to be treated as a single electrode system in this sense.

A metal underground gas piping system shall not be used as a grounding electrode.

Rod and pipe electrodes shall not be less than 8 feet (2.44 m) in length and may consist of the following materials:

Electrodes of pipe or conduit shall not be smaller than 3/4-inch trade size and, where of iron or steel, shall have the outer surface galvanized or otherwise metal-coated for corrosion protection.

Electrodes of rods of iron or steel shall be at least 5/8 inch (15.87 mm) in diameter. Nonferrous or stainless steel rods or their equivalent less than 5/8 inch (15.87 mm) in diameter shall be listed and shall be not less than 1/2 inch (12.7 mm) in diameter.

The electrode shall be installed such that 8 feet (2.44 m) of length is in contact with the soil. It shall be driven to a depth of not less than 8 feet (2.44 m) except that where rock bottom is encountered, the electrode shall be driven at an oblique angle not to exceed 45 degrees from the vertical or shall be buried in a trench that is at least 2 1/2 feet (762 mm) deep. The upper end of the electrode shall be flush with or below ground level unless the aboveground end and the grounding electrode conductor attachment are protected against physical damage.

Plate Electrodes. A plate electrode must expose not less than 2 square feet (0.186 sq m) of surface to exterior soil. Electrodes of iron or steel plates shall be at least 1/4 inch (6.35 mm) in thickness. Electrodes of nonferrous metal shall be at least 0.06 inch (1.52 mm) in thickness.

Each electrode shall be separated by at least 6 feet from other electrodes used for signal circuits, radio, lightning rods, television, and other purposes. Although not in the *Code*, it is considered good practice to bond the electrodes together. In fact, doing this will solve many problems.

Made electrodes must have a resistance to ground of 25 Ω (ohms) or less, wherever practicable. When the resistance is greater than 25 Ω, two or more electrodes may be connected in parallel or extended to a greater length. The *Code* does not always go into the mechanics of grounding, but good practice is that the electrode has a lower resistance when driven some distance from a foundation into undisturbed soil, where the earth will put pressure on the driven electrode.

Water piping usually has a resistance of 3 Ω or less. Metal frames of buildings often make a good ground, especially where they contact rebar in the concrete, and usually have a resistance of less than 25 Ω. As was pointed out in *Section 250.52*, the metal frame of a building (when effectively grounded) may be used as the ground. Local metallic water systems and well casings also make good grounds in most cases. One might wonder why metal frames are covered in dealing with residences, but there is a trend to new types of residential construction, and one of these uses metal studs.

Grounding, when especially difficult due to poor soil conditions, can be greatly improved by the use of chemicals such as magnesium sulfate, copper sulfate, or rock salt. Note, however, that these chemicals will tend to corrode the grounding electrode, making it measurably less effective in only a few years. A doughnut-type hole may be dug around the ground rod and the chemicals put into the hole. Another method is to bury a tile close to the rod and fill the tile with the chemical. Rain and snow will dissolve the chemicals and allow them to penetrate the soil, lowering its resistance (Figure 2-32).

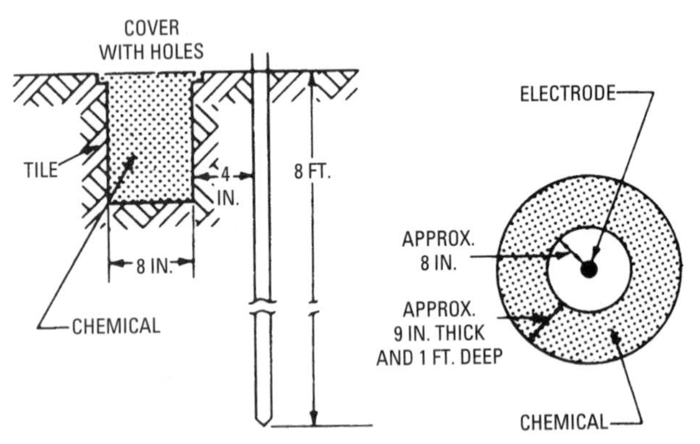

Figure 2-32 Adding chemicals to the soil to lower its ground resistance.

The *NEC* recommends that the resistance of the grounds be tested periodically after installation. This is rarely done. In fact, it is practically never done even at the time of installation, except by utility companies that realize the importance of an adequate ground. The testing of grounds is a mystery to many electricians. Do not try to use a common ohmmeter for the testing—the readings obtained are apt to be almost anything, due to stray AC or DC currents in the soil or due to DC currents set up by electrolysis in the soil. If you must perform this test, specialized ground testers are available. An example of what might be expected by paralleling ground rods follows (these figures are general and should not be taken as the results in every case): Two rods in parallel, with a 5-foot spacing between them, will reduce the resistance to about 65 percent of what one rod would be. Three rods parallel with the same spacing will reduce the distance to about 42 percent; while four rods paralleled will reduce the resistance to about 30 percent.

Grounding Conductors The *Code* regulates the material and installation of grounding electrode conductors and equipment grounding conductors. The material for grounding conductors should be as follows:

Grounding Electrode Conductors. This is the conductor that runs from the electrical service to the grounding electrode. It may be:

1. Copper or other corrosive-resistant material
2. Solid or stranded
3. Insulated, covered or bare
4. Without splice or joints, except in the case of bus bars or if the splice is made with an exothermic weld or an irreversible compresion connector.

Electrical resistance per foot (linear) shall not exceed that of the allowable copper conductors that might be used for this purpose. Thus, if aluminum (in cases where permissible) is used, the conductor will have to be larger than copper would be for the same purpose.

Equipment Grounding Conductors. This is the (green) grounding conductor for equipment, conduit, and other metal raceways or conductor enclosures, *not* the grounding electrode conductor. It may be:

1. Copper or other corrosive-resistant material
2. Bus bar, rigid conduit, steel pipe, electrical metallic tubing (EMT), the armor of AC metal-clad cable, or intermediate metal conduit
3. Stranded or solid

Joints of conduit must be made up wrench-tight, not plier-tight.

It is well to state at this point that, where 10 feet or more of buried metallic water piping are used as the common grounding electrode (*Section 250.52*), and where a water meter is present, either the common grounding conductor must be connected on the street side of the water meter or, if connected on the house side, the water meter will have to be bonded. This clause will no doubt be interpreted to mean the use of valves, unions, or other connectors (Figure 2-33).

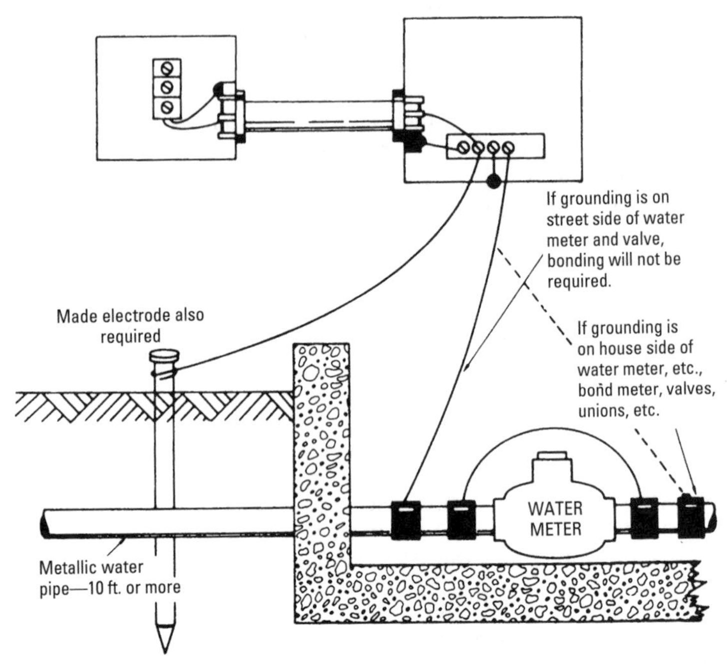

Figure 2-33 Proper bonding of a water meter.

Grounding conductors shall be installed as follows:

Grounding Electrode Conductors. No. 4 or larger conductors may be attached to the surface—knobs or insulators are not required. Mechanical protection will be required only where the conductor is subject to severe physical damage. No. 6 grounding conductors may be run on the surface of a building if protected from physical damage and rigidly stapled to the building structure. Grounding conductors smaller than No. 6 shall be in conduit, EMT, or armor. One might just as well forget No. 8 copper as a common grounding conductor unless it is run in a raceway.

The metallic enclosures for the grounding conductor shall be continuous from the cabinet to the grounding electrode and shall be attached at both ends by approved methods (Figure 2-34). If the conduit or other raceway, is used for protection purposes only, then the common grounding conductor shall be bonded to the metallic raceway at one or both ends as required (Figure 2-35). Caution: Do not confuse the common grounding (neutral) conductor, which we are discussing here, with the equipment-grounding conductors.

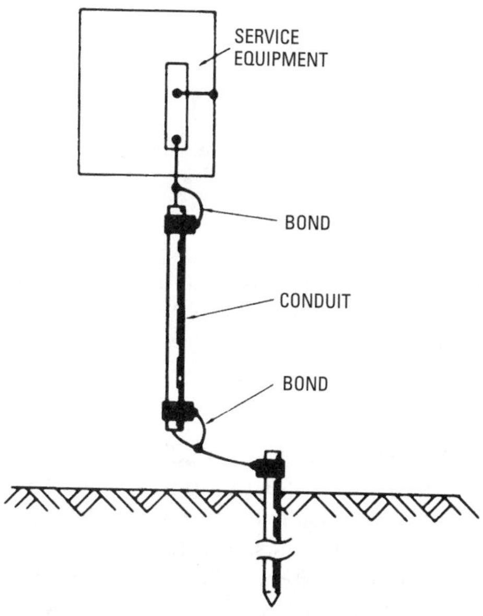

Figure 2-34 Conduit or armor used to protect the grounding wire shall be bonded to the grounding electrode and to the service-entrance equipment.

Because of corrosion, aluminum grounding conductors shall not be placed in direct contact with masonry, earth, or other corrosive materials. Also, wherever aluminum grounding conductors are used, they shall not be closer than 18 inches to the earth. Please note that this does not prohibit the use of aluminum grounding conductors, but the restrictions placed upon them will practically eliminate their use, since they are not to be spliced.

Table 250.66 (*Table 2-2 in this book*) is used for sizing the grounding electrode conductors in grounded systems.

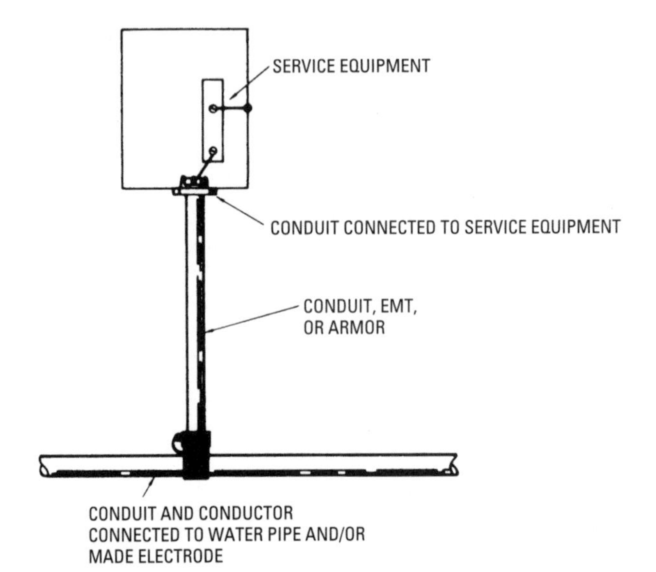

Figure 2-35 Where the conduit or armor is used for protection only, the common grounding conductor shall be bonded to the metallic raceway at one or both ends, as required.

Grounding Conductor Connections. Care shall be taken to ground raceways or cable armor or interior wiring by connecting to grounding conductors as near as possible to the source of supply. Also, the grounding conductor shall be chosen so that no raceway or cable armor is grounded by a grounding conductor smaller than that required by *Table 250.122*. Refer to the *NEC* for this table.

Connections to the Grounding Electrode. The grounding connection to the electrode shall be located as follows:

To Water Pipes. Grounding electrode conductors shall be attached to water pipes:

1. On the street side of the water meter
2. To a cold-water pipe of adequate current-carrying capacity
3. As near as possible to the entrance of the water supply into the building

Table 2-2 Grounding Electrode Conductors for AC Systems

Size of Largest Service-Entrance Conductor or Equivalent Area for Parallel Conductors		*Size of Grounding Electrode Conductor*	
Copper	***Aluminum or Copper-Clad Aluminum***	***Copper***	****Aluminum or Copper-Clad Aluminum***
2 or smaller	0 or smaller	8	6
1 or 0	2/0 or 3/0	6	4
2/0 or 3/0	4/0 or 250 kcmil	4	2
Over 3/0 through 350 kcmil	Over 250 kcmil through 500 kcmil	2	0
Over 350 kcmil through 600 kcmil	Over 500 kcmil through 900 kcmil	0	3/0
Over 600 kcmil through 1100 kcmil	Over 900 kcmil through 1750 kcmil	2/0	4/0
Over 1100 kcmil	Over 1750 kcmil	3/0	250 kcmil

*Where there are no service-entrance conductors, the grounding electrode conductor size shall be determined by the equivalent size of the largest service-entrance conductor required for the load to be served.

If not on the street side of the water meter, the meter shall be adequately bonded and the bonding jumper shall be long enough so that the water meter can be removed without disturbing the jumper (Figure 2-33).

Care must be taken to place grounding conductors on the street or supply side of insulated couplings, unions, valves, and other connectors.

Attachment to water pipes shall be accessible if at all possible.

If the water supply is from a well on the premises, the point of attachment shall be as close as possible to the well, with the casing bonded to the water piping (Figure 2-29).

Connections to Other Electrodes. Grounding to electrodes other than metallic water piping is permitted in *Section 250.52*. When grounding to this type of electrode, make certain that it is a permanent ground and that the connection is accessible if at all possible.

Chapter 3

Branch Circuits and Loads

Branch circuits comprise the vast majority of the wiring in a house. These are relatively small circuits, but there is a large number of them. Twenty branch circuits in a single home would be fairly common, and thirty to forty in a larger home would be fairly common as well.

Once the electrical service to a house is designed, the rest of the design process focuses primarily on branch circuits. We will cover the materials and methods pertaining to branch circuits in Chapter 4. In this chapter, we are addressing the loads placed on these circuits.

Bear in mind that the loads placed on these circuits will change continually. (An electrical *load* is any current-using device. A radio is a load, a lighting fixture is a load, and so on.) In a house, loads are plugged into receptacles and turned on and off as the need arises. Thus loading of circuits may change from year to year or even from day to day. It is important to design circuits that will be sufficient to carry the loads that will be placed upon them.

Certainly, if too many loads are placed on a single circuit, its *overcurrent protective device* (fuse or circuit breaker) will trip and disconnect the circuit from the power source. But this is not the preferred method for preventing excessive current draw. Ideally, the circuits should be laid out in such a way that not too many loads would be placed upon them in the first place. It is done by limiting the number of outlets connected to any single circuit. In most electrical design, no more than eight outlets are connected to a 20-ampere circuit. This is not a strict requirement of the *NEC*, but it is generally considered good practice. All outlets are included in this figure—not just receptacle outlets, but lighting outlets as well.

For areas where larger loads may be expected (such as a kitchen, where portable appliances will be used), fewer outlets are connected to any single circuit.

Installing some extra cable and a few extra circuit breakers is the only cost associated with adding circuits during the time of construction. This is a very small cost. Once the house is completed, however, adding new circuits involves adding new cables through existing finished walls. This is a very difficult and expensive process. Good trade practice is to provide more than enough circuits at the beginning.

Definitions

In order to plan the individual circuits within a house, it is necessary to do a certain amount of calculating for branch circuits (as well as for the electrical service and, occasionally, feeder circuits). The definitions of these terms are as follows:

Branch Circuits

There are five separate definitions for branch circuits, each with an individual purpose in mind. Only four out of five definitions are given here as it is not felt that the other one particularly applies to residential wiring.

1. A branch circuit is that portion of a wiring system extending beyond the final overcurrent device protecting the circuit. This protective device is a circuit breaker or fuse in the service-entrance equipment. (Circuit breakers or fuses for feeder circuits to the feeder panels are not branch circuit protective devices.) The branch circuit is that circuit coming from the service-entrance equipment, or in some cases, from a feeder panel.

   The phrase "protecting the circuit" defines what is meant by "overcurrent protective device." Thus, in the lighting branch circuits shown in Figure 3-1, there is a circuit breaker or fuse in the panel feeding the lighting branch circuit.
2. A branch circuit, as applied to appliances, is a circuit designed for the purpose of supplying one or more specific appliances;

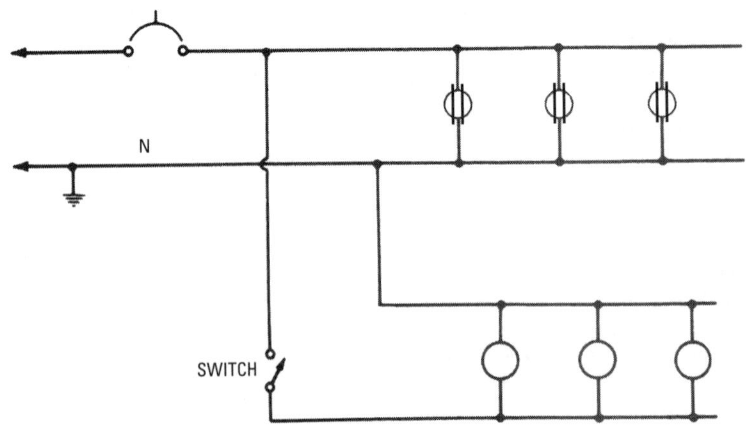

Figure 3-1 A lighting branch circuit.

nothing else can be connected to this circuit, including lighting. (The lighting that is an integral part of the appliance is not considered as lighting in this instance.) Pay particular attention to this part, as we will be covering appliance circuits as they apply to residential wiring.

3. A multiwire branch circuit (Figure 3-2, parts A and B) is a circuit consisting of two or more ungrounded conductors with an equal potential between them, and a grounded conductor with an equal potential between it and any one ungrounded conductor. In residential wiring, this is a 120/240-volt system in most cases, but it may also be a 208Y/120-volt system. The circuit shown in Figure 3-2, part C, is not a multiwire branch circuit, because it utilizes two ungrounded conductors from the same phase in conjunction with the neutral conductor.
4. A branch-circuit individual is a circuit that supplies just one piece of equipment, such as a motor, an air conditioner, or a furnace.

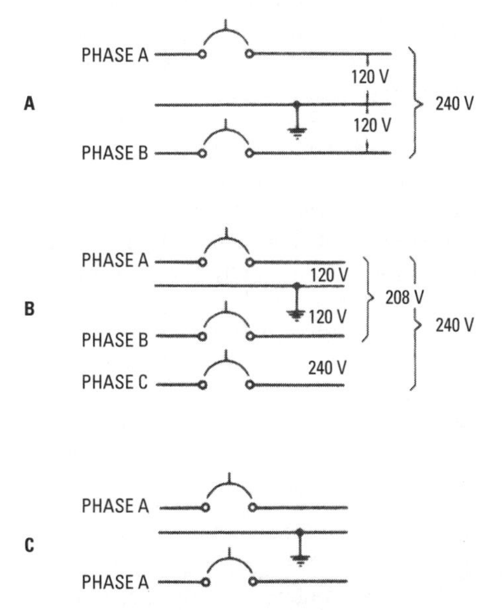

Figure 3-2 Variations of a multiwire branch circuit. Circuit C is not a multiwire branch circuit, because it utilizes two wires from the same phase in conjunction with the neutral conductor.

Feeder

Feeders are all circuit conductors between the service equipment (or the generator switchboard of an isolated plant) and the final branch-circuit overcurrent device. We must always have service equipment, but it is not always necessary to have feeders. For instance, the average home does not have feeders; the branch circuits are taken from the overload devices in the service-entrance equipment. When we have a large area to cover, the usual practice is to extend feeder circuits from the service-entrance equipment to the proper locations for distribution to the branch circuits. Thus, the conductors from the service equipment to the distribution location are termed feeders.

Calculations

In the first part of this chapter, the loading of circuits was discussed. The subject of this section, however, is determining the minimum sizes and number of such circuits.

The minimum size of circuits is determined by the *NEC* or the local electrical inspector. The *Code* specifies a minimum circuit size of 15 amperes, using No. 14 AWG copper wire. (A few minor exceptions exist that are not applicable to house wiring.) However, many electrical inspectors prefer a minimum of a 20-ampere circuit, using No. 12 AWG wire. In fact, 20 amperes on No. 12 conductors has become almost a standard in recent years.

To determine the number of circuits required for a house, however, special calculations must be performed. (Sometimes these calculations must be presented to an electrical inspector or in order to obtain a permit, but not always.) We will go through the specific calculations in the following paragraphs. They are specified by the *NEC*. As mentioned previously, the requirements of the *NEC* are based upon many decades of experience, trial, error, and corrections. These rules must be followed if safety is to be ensured.

It is important to stress that these are minimums. In actual practice, significantly more than the minimum is considered a good practice. Modern homes use far more power and circuits than the homes of even a few decades ago, so going 25 to 50 percent beyond the minimums is sensible.

General Lighting Loads

For the information necessary to arrive at the calculations of loads, we will use 3 watts per square foot for residential occupancies. There will be some cases where this figure will not be sufficient, such as when more lighting than normal is installed in a residence. For general calculations, however, 3 watts per square foot will be adequate.

Should there be extra lighting, this will be added to the total obtained from the 3-watts-per-square-foot figure.

In figuring the watts per square foot for a residence, the outside dimensions of the building shall be used. These dimensions do not include the area of open porches or garages that are attached to the residence. If there is an unfinished basement, it should be assumed that it will be finished at a later date. Thus, the square-foot area of the basement should be included in the square-foot area used for load calculations so that the system will be adequate when the basement is finished.

In Figure 3-3, part A, we find that the first floor measures 32½ feet by 57 feet, or 1,852½ square feet. The basement, shown in Figure 3-3, part B, is only a partial basement with measurements of 32½ feet by 34 feet, or 1,105 square feet. From this we find a total of 2,957½ square feet, which will be used with the 3-watts-per-square-foot figure for general lighting. We find that this comes to a total of 8,872½ watts.

This number of watts per square foot is a general lighting load. Any outlet of 15 amperes or less, such as receptacles in bedrooms, living rooms, or bathrooms, is considered in the general lighting load. Any loads other than those considered as general lighting loads, such as special lighting, heating, air cooling (a special loading is considered for heating and air cooling, and will be covered later), or any special motor loads, will be considered separately. Ranges and electric clothes dryers will also be considered separately.

In general, for illumination as covered under watts per square foot in dwelling occupancies, it is recommended that not less than one branch circuit be installed for each 500 square feet of floor area in addition to that required for special loads.

Section 210.11(C)(1) requires what are termed *small appliance branch circuits*. These are not to include any appliances that are fixed, such as disposals or built-in dishwashers. The specific requirements are expressed by *Section 210.52(B)(1)*. These sections require that to supply the small appliance load in the kitchen, pantry, family room, dining room, and breakfast room of dwelling occupancies, two or more 20-ampere branch circuits in addition to other branch circuits must be provided for all receptacle outlets in these rooms, and such circuits shall have no other outlets. Receptacle outlets supplied by at least two appliance receptacle branch circuits shall be installed in the kitchen.

Section 210.11(C)(2) calls for at least one 20-ampere branch circuit to be provided for laundry receptacles. Details are provided in *Section 210.52(F)*.

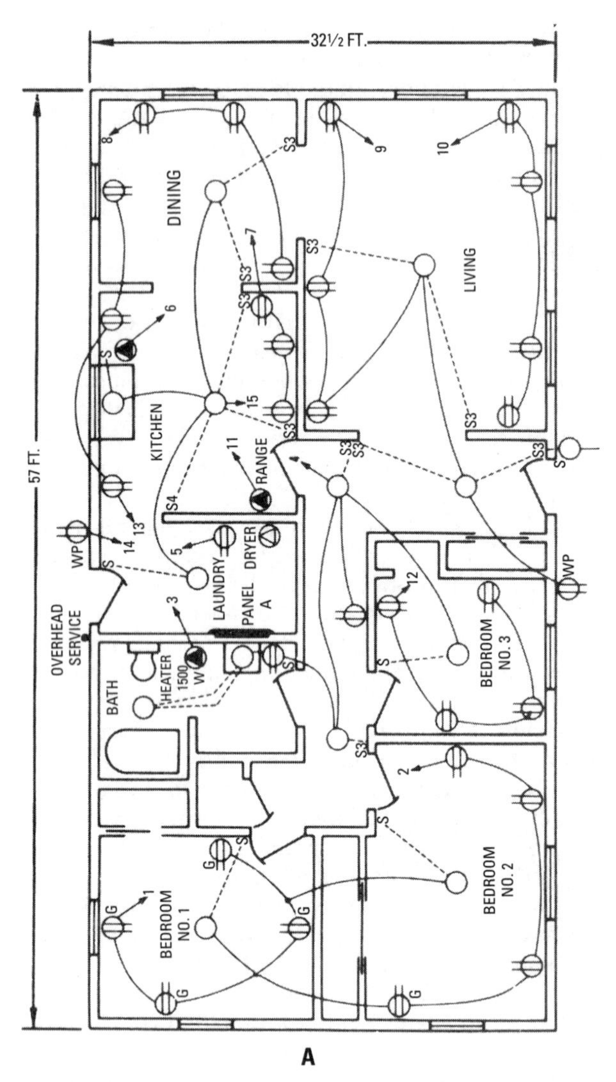

A

Figure 3-3 Electrical layout drawing of a one-story residence with basement. (A) First-floor layout.

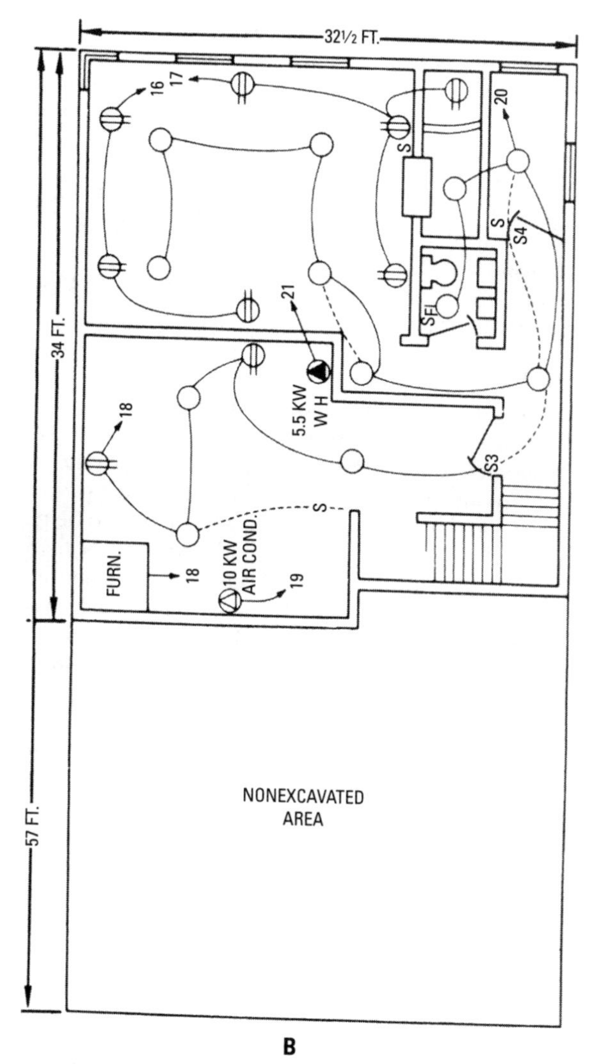

B

Figure 3-3 (*continued*) (B) Basement layout.

Additions to Existing Installations

While most of our coverage is of wiring new houses, additions to existing homes are also common types of house wiring. When adding to an existing installation, a special calculation must be done. New circuits or extensions to existing electrical systems in dwellings are calculated on the watts-per-square-foot basis or the amperes-per-outlet basis. This will apply to that portion of an existing building that has not been previously wired or to any addition that exceeds 500 square feet in area. In this case, the addition will be figured on the watts-per-square-foot basis, per *Section 220.3(C)*. Also take into consideration any other loads that are involved.

Small Appliance Loads

Every house is required to have two small appliance branch circuits. These are traditionally run to a refrigerator outlet in the kitchen and to a receptacle outlet above a kitchen counter. The location of this countertop outlet is calculated to be where the largest loads are likely to be plugged in. These appliances are likely to be microwave ovens, food processors, hot plates, or other relatively high-wattage devices. The small appliance branch circuits just mentioned shall have a feeder load of not less than 1,500 watts for each two-wire circuit installed, as outlined in *Section 220.16(A)*. These circuits are for small appliances only (portable appliances supplied from receptacles of 15- or 20-ampere rating). Recall that a minimum of two such 20-ampere (1,500-watt) circuits shall be installed in the kitchen. It will often be desirable to install more than the required minimum of two. Remember that small appliance circuits are also required in the pantry, dining room, and breakfast room. These may be fed from the two required in the kitchen, or separate 1,500-watt (20-ampere) circuits may be installed. These 20-ampere small appliance circuits and the 20-ampere laundry circuit must be installed with No. 12 copper conductors or No. 10 aluminum conductors.

Electric Ranges

Electric ranges are frequently the largest single load in a house. The most common rating for a range circuit is 50 amperes at 240 volts. Because this is such a large load, and because the device is used intermittently (How often will all the burners and the oven all be on at the same time?), the *NEC* provides a separate section dealing with the calculation of range loads. In calculating feeder loads for electric ranges or other cooking appliances in dwelling occupancies, any that are rated over $1^3/_4$ kW shall be calculated according to *Table*

3-1 (*Table 220.19* in the *NEC*). Some of the notes accompanying the table in the *NEC* are a part thereof and are important in the calculation of feeder and branch circuits.

Because of the larger wattages being used in modern electric ranges, it is recommended that the minimum demands for any range with less than 8¾-kW rating be figured using Column A in *Table 3-1*. The notes accompanying this table do not appear here. Should you need to refer to them, see the *Code*.

When using this table, it is wise not merely to size the branch circuit and/or feeder circuit to the cooking unit(s) installed according to this table but to have additional capacity in the conductors so that if a higher-wattage range is added later, it will not be necessary to rewire the circuit. The additional cost at the time of the original installation will be very small.

The demand factor, as applied in the table, may be applied to an entire electrical system or to any part of an electrical system. It is the ratio of the maximum demand of a system, or part of a system, to the total connected load of a system or the part of the system under consideration. The loads on a system are practically never thrown on at the same time due to the diversity of uses. Somewhere between the maximum connected load and the actual usage is a load that may be considered the maximum demand. This fact is often used in determining the size of conductors or overcurrent devices. The demand factor is usually determined by a series of tests and is added to the *Code* after it is proved. Examples will appear later.

As a rule, the entire wattage of the range will not be used at one time, and this is the reason why demand factors have been taken into consideration in *Table 3-1*.

Clothes Dryers

Electric clothes dryers typically require a 30-ampere, 240-volt circuit, making it another large load in a house. (Gas clothes dryers typically require only a 15- or 20-ampere, 120 volt circuit.) Electric clothes dryers will be taken at 100 percent of the nameplate rating. As with ranges, it is a good policy to install the branch circuit and feeder conductors a little larger than the minimum required. The author recommends nothing less than No. 8 copper or No. 6 aluminum to the dryer. Some wirers have a tendency to install conductors that barely meet the nameplate rating, but with higher-wattage driers appearing, it certainly is false economy not to install sufficiently large conductors in the original installation.

Table 3-1 Demand Loads for Household Electric Ranges, Wall-Mounted Ovens, Counter-Mounted Cooking Units, and Other Household Cooking Appliances over $1^{3}/_{4}$ kW Rating

	Maximum Demand (kW)	*Demand Factors (Percent)*	
Number of Appliances	*Column A (Not Over 12 kW Rating)*	*Column B (Less Than $3^{1}/_{2}$ kW Rating)*	*Column C ($3^{1}/_{2}$ kW to $8^{3}/_{4}$ kW Rating)*
1	8	80	80
2	11	75	65
3	14	70	55
4	17	66	50
5	20	62	45
6	21	59	43
7	22	56	40
8	23	53	36
9	24	51	35
10	25	49	34
11	26	47	32
12	27	45	32
13	28	43	32
14	29	41	32
15	30	40	32
16	31	39	28
17	32	38	28
18	33	37	28
19	34	36	28
20	35	35	28
21	36	34	26
22	37	33	26
23	38	32	26
24	39	31	26
25	40	30	26
26–30	{15 plus 1 kW for each	30	24
31–40	range}	30	22
41–50	{25 plus $^{3}/_{4}$ kW for	30	20
51–60	each range}	30	18
61 and over		30	16

Space Heating and Cooling

Heating and cooling are obviously among the most important mechanical systems for a house. Cooling (air conditioning, AC) is almost always electrically powered (a few gas-powered AC systems do exist, however). Air conditioning circuits are usually 30 amperes at 240 volts.

Heating systems may be electrically-powered, oil-burning, or powered with natural gas. Heating units using oil or gas require only small circuits, typically 15 or 20 amperes at 120 volts. Electric heating circuits, on the other hand, are usually 50-ampere circuits operating at 240 volts. In dwelling occupancies having both electrical space-heating and air-cooling equipment, the larger of the two loads is used in calculations and the smaller is omitted, providing that the likelihood of both being used at the same time is remote. (Heating in the winter; air conditioning in the summer.) An air-conditioning load is an inductive-type load, so the number of amperes drawn by this equipment is used in the calculations in order to take into account the low power factor involved.

Farm Buildings

Wiring on farms is allowed by the *NEC* to differ slightly from wiring in other types of residences. There is a subset of rules addressing only wiring on farms, and on the various buildings on farms. Calculations for farm buildings will appear in Appendix B.

Calculations of Feeder Loads

Feeders are uncommon in most homes. A large home may have a 60- or 100-ampere feeder running from the main service to a large area of the home or to a large outbuilding. In such cases, the 3-wire feeder (with ground) will run from a single two-pole circuit breaker in the main panel to a smaller circuit breaker near the loads fed by the feeder.

A more common instance of feeders in residential wiring is for mobile homes. This application will be covered in detail in Chapter 6.

Table 3-2 (*Table 220.11* in the *Code*) applies to dwellings.

Motor Loads

Motors are unusual loads for several reasons. First of all, motors are subject to high starting currents. Such current levels can easily be four to eight times the motor's normal operating current. Secondly, motors may draw unusual levels of current if operating improperly.

Table 3-2 Lighting Load Feeder Demand Factors

Type of Occupancy	*Portion of Lighting Load to Which Demand Factor Applies (volt-amperes)*	*Demand Factor Percent*
Dwelling units	First 3,000 or less at	100
	From 3,001 to 120,000 at	35
	Remainder over 120,000 at	25
*Hospitals	First 50,000 or less at	40
	Remainder over 50,000 at	20
*Hotels and motels—including apartment houses without provision for cooking by tenants	First 20,000 or less at	50
	From 20,001 to 100,000 at	40
	Remainder over 100,000 at	30
Warehouses (storage)	First 12,500 or less at	100
	Remainder over 12,500 at	50
All others	Total	100

*The demand factors of this table shall not apply to the computed load of feeders to areas in hospitals, hotels, and motels where the entire lighting is likely to be used at one time, as in operating rooms, ballrooms, or dining rooms.

> *220.14. Motors. Motor loads shall be computed in accordance with Sections 430.24, 430.25, 430.26, and Section 440.6 for hermetic refrigerant motors and compressors.*

The *NEC* states here that if the motor loads are included in the residence (and this will include air-conditioning motors, furnace motors, pump motors, disposal motors, and so forth), 125 percent of the full-load current rating of the largest motor involved will be taken, and then 100 percent of the full-load current rating of the balance of the motors used in determining the ampere rating of the conductors required for the service and/or feeders in the residence.

Placing Outlets

In general, space outlets according to the following guidelines (equally apart, as far from one another as possible). Even if you vary from this model, do not leave more than 12 feet between each outlet. Twelve-foot spacing allows lamps with their usual 6-foot-long cords to be placed where needed and still reach an outlet (Figure 3-4).

1. General-purpose outlets should be placed within 6 feet of lamps, television sets, and other small appliances.
2. There should be one outlet at least every 12 feet on every wall.

3. Outlets for general use should be about 12 inches off the floor.
4. Switches should be 4 feet from the floor and within 6 inches of doors and archways.
5. Kitchen receptacles are placed about 12 inches above countertops, or 4 feet off the floor.
6. There should be a kitchen outlet at least every 4 feet in the work area.
7. Outdoor wiring and bathroom outlets must be channeled through a groundfault circuit interrupter (GFCI).
8. Consider fluorescent fixtures for softer, less energy-consuming light.

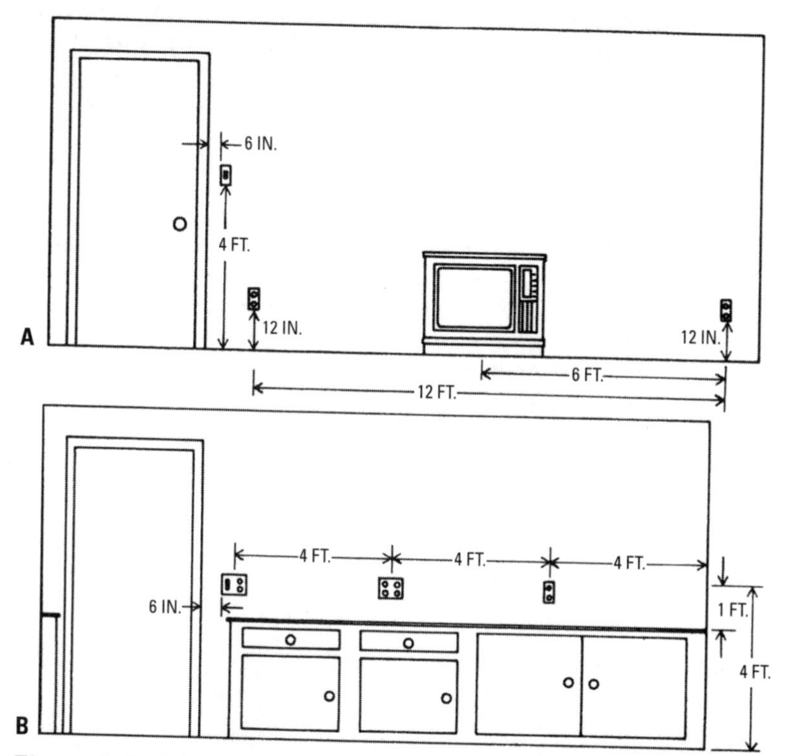

Figure 3-4 (A) Maximum distances for placement of general outlets and switches and (B) The *NEC* requirements for placement of outlets and switches in the kitchen.

When wiring a family room, remember that there will probably be a television set, perhaps with a rotary antenna, requiring two outlets at one spot. There may also be a stereo, room air conditioner, or other power users in the same area, so space outlets close together or consider multiple outlets in one spot (which requires two or more ganged boxes). Be generous with your outlets when putting in new work. It is simpler, easier, and less expensive to do it now than later.

Use similar foresight in all your planning. Don't skimp on outlets just because you cannot see a need for them in the immediate future. Twenty years ago, no one could predict a home with multiple TVs and stereos, videotape recorders, individual hair dryers and hair setters, microwave ovens, and the other electronic marvels found in our homes today.

Placing Switches

Wall switches are another often neglected planning aspect. Just because there are no wall or ceiling lights does not mean that there is no need for switches. Even when all lamps are the plug-in type, switches near room entrances are always a good idea. These switches can control one or more outlets in the room and save you from fumbling in the dark. In a bedroom, for example, connect a doorway switch to an outlet near a dresser or vanity. This way, a person can flick on the light as soon as he or she enters the room, instead of stumbling about trying to find the lamp. In many parts of the United States, codes require certain outlets to be connected to a wall switch by a door. Use the same reasoning for outdoor lights. An indoor switch can turn on patio or driveway lights and prevent falls over a garbage can or down the steps.

Chapter 4

Materials and Methods

All electrical systems are constructed using specific types of materials, which are designed to be put together according to certain methods. We generally call these materials and methods *wiring systems*.

In this chapter, we will identify the wiring methods that are used in house wiring and identify the requirements for their proper application.

Nonmetallic-Sheathed Cable (Types NM and NMC)

Nonmetallic-sheathed cable, as covered in *Article 334* of the *NEC*, is very commonly used in the wiring of residences. It is probably the most widely used type of wiring for this purpose. NM cable is often referred to by a brand name such as Romex. These are nonmetallic-sheathed cables approved in sizes No. 14 through No. 2 AWG with copper conductors and in sizes No. 12 through No. 2 AWG with aluminum conductors (by conductors meaning the current-carrying conductors in these cables). In addition to the two or three current-carrying conductors (Figure 4-1), the cable will also contain an equipment-grounding conductor of proper size, as specified in *Table 250.122* in the *Code*. This grounding conductor may be bare or covered with green insulation, or green insulation with one or more yellow stripes. It is definitely not a current-carrying conductor but is for grounding receptacle boxes, receptacles, switch boxes, and equipment of any nature, including enclosures, raceways, and appliances. This grounding conductor is an equipment-grounding conductor, and the only time that it will carry current is when a fault develops in the circuit.

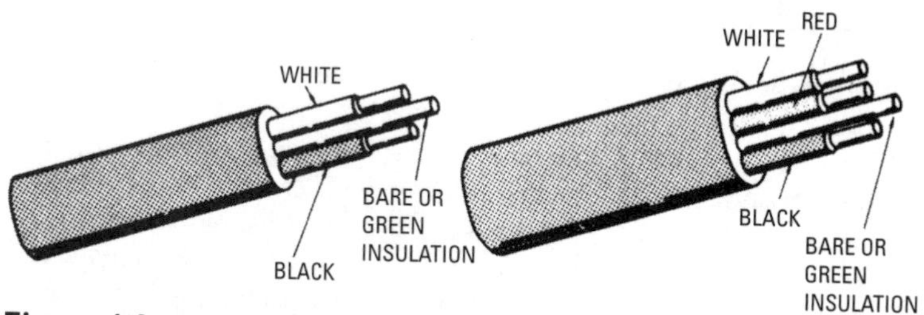

Figure 4-1 NM, NMC, or UF cable.

When the grounding-type receptacle appeared in the *NEC* (shown here in Figure 4-2), there were as yet no current-using devices on the market that had cords including an equipment-grounding conductor and the grounding-type attachment plug (the latter shown in Figure 4-3). Along with this came NM cable with ground for grounding purposes. Since the appearance of the grounding-type receptacles, a sufficient number of this type have been installed on grounded circuits that it became practical to make some changes in the *Code*—mainly to require the grounding of more appliances than were required to be grounded under previous *Codes*.

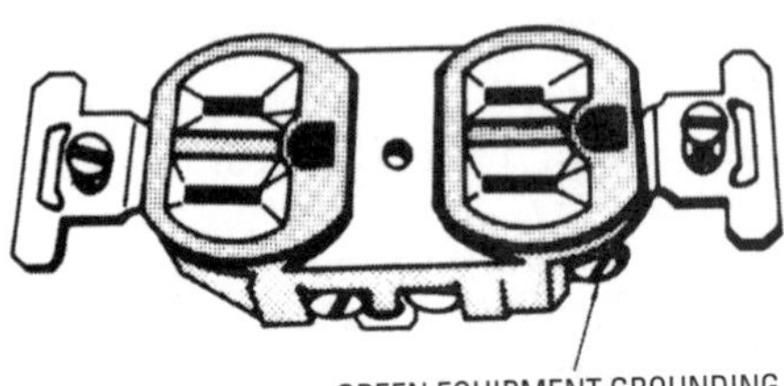

Figure 4-2 A grounding-type receptacle.

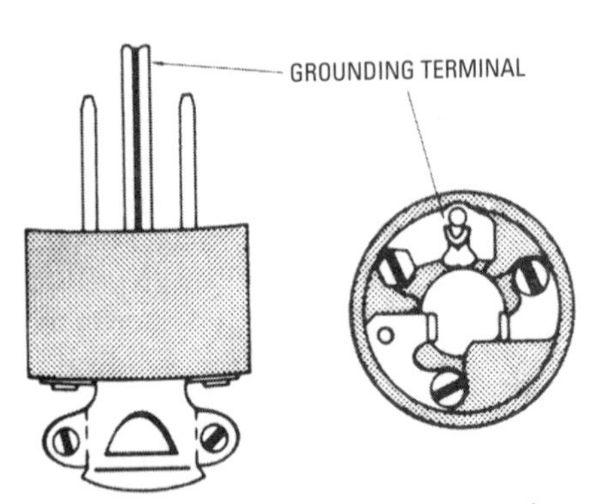

Figure 4-3 A grounding-type attachment plug.

Prior to the appearance of this type of equipment grounding, it was necessary on certain appliances, such as an electric washing machine, to ground the appliance frame by means of a separate grounding conductor clamped to the cold-water piping. This was a rather ineffective grounding method, because high impedances were often encountered.

Now, with the grounding-type receptacles, NM cable with ground, and the spelling out of how the grounding is to be accomplished, we have a very effective equipment-grounding circuit that, when properly installed, makes for a low impedance and thus for a more effective means of operating the overcurrent devices or devices protecting the circuit in case of ground faults. A word of caution at this point: In wiring existing residences, you shall definitely not install a grounding-type receptacle on a circuit that does not have an equipment ground. This would create false security, and accidents could follow. The *NEC* has taken into account that we will run into existing ungrounded circuits on which extensions for additional receptacles must be made. For extensions only in existing installations that do not have a grounding conductor in the branch circuit, the grounding conductor of a grounding-type receptacle outlet may be grounded to a grounded cold-water pipe near the equipment as shown in Figure 4-4. *Caution:* This does not make the entire circuit a grounded circuit; it merely grounds the receptacles added to an existing circuit.

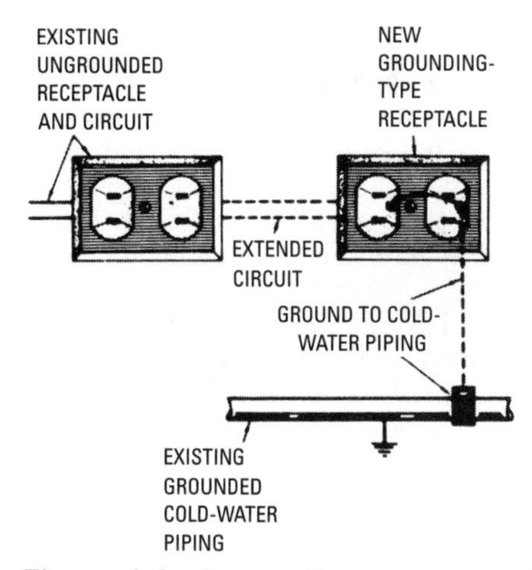

Figure 4-4 A grounding-type receptacle grounded to a cold-water pipe when installed in an existing undergrounded system.

Another important subject is covered in *Section 250.114, Equipment Connected by Cord and Plug.* In residential occupancies, the following exposed noncurrent-carrying metal parts of cord- and plug-connected equipment, which are liable to become energized, have to be grounded:

> *(1) refrigerators, freezers, and air conditioners; (2) clothes-washing, clothes-drying, dishwashing machines, sump pumps, electrical aquarium equipment; (3) handheld motor-operated tools; (4) motor-operated appliances of the following types: hedge clippers, lawn mowers, snow blowers, and wet scrubbers; (5) portable handlamps.*

The exception to this list is that listed equipment protected by a system of double insulation, or the equivalent, is not required to be grounded. Where such a system is employed, the equipment must be distinctively marked.

A portion of *Table 250.122* in the *Code*, which gives us the sizing for equipment grounding conductors, is reproduced here as *Table 4-1*. Please note that for 15-, 20-, and 30-ampere circuits, the equipment grounding conductors are the same size as the phase conductors. Larger circuits permit a reduction in the size of the equipment grounding conductors.

Use and Installation of NM and NMC Cables

Nonmetallic sheathed cable is generally considered the easiest type of cable to install, and because of this, it is preferred by many contractors. But although it is generally a fine wiring method, Type NM cables cannot be installed everywhere. Although NM cable is permitted in almost all houses, knowing where it can and cannot be installed is very important.

Section 334.10 gives us the information where nonmetallic-sheathed cable may be used. Fundamentally, it may be used for both exposed and concealed installations. The following are *not permissible* uses for *either* NM or NMC cables:

1. As service-entrance cable
2. In commercial garages
3. In theaters, except as provided in *Article 518, Places of Assembly*
4. In motion picture studios
5. In storage battery rooms
6. In hoistways
7. In any hazardous location

Table 4-1 Minimum Sizes of Equipment-Grounding Conductors for Grounding Raceway and Equipment

Rating or Setting of Automatic Overcurrent Device in Circuit Ahead of Equipment, Conduit, etc., Not Exceeding (Amperes)	*Size*	
	Copper Wire No.	*Aluminum Wire No.**
15	14	12
20	12	10
30	10	8
40	10	8
60	10	8
100	8	6
200	6	4
400	3	1
600	1	2/0
800	0	3/0
1000	2/0	4/0
1200	3/0	250 kcmil
1600	4/0	350 kcmil
2000	250 kcmil	400 kcmil
2500	350 kcmil	600 kcmil
3000	400 kcmil	600 kcmil
4000	500 kcmil	800 kcmil
5000	700 kcmil	1200 kcmil
6000	800 kcmil	1200 kcmil

*See installation restrictions in Section 250. 92(A).

8. Embedded in poured cement, concrete, or aggregate
9. In any structure exceeding three floors above grade

Type NM cable, as previously stated, may be installed for both exposed and concealed work in normally dry locations. This would include installing or fishing of NM cable in the air voids in masonry block or tile walls, but only if such walls are not exposed to excessive moisture or dampness. An example of Type NM cable installation in masonry blocks would be in blocks used in the interior walls of buildings. Exterior walls of block, including exterior basement walls, shall not have NM cable installed in the voids of these blocks (Figure 4-5).

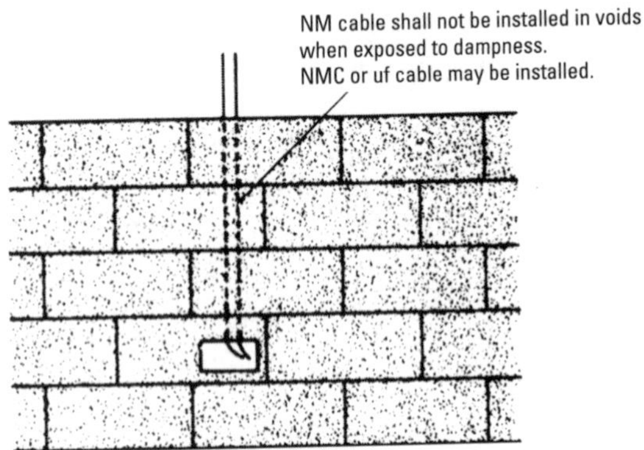

Figure 4-5 NMC or UF cable may be installed in the voids of concrete-block walls, but not NM cable.

NMC cable, in contrast to NM cable, is a moisture- and corrosion-resistant cable. Therefore, NMC cable may be installed for both exposed and concealed work in any place that NM cable would be permitted, and it may also be installed in moist, damp, or corrosive locations. Thus, it may be installed in the hollow voids of masonry block or tile walls used as outside walls of buildings, both above-ground walls and basement walls. NMC cable may also be embedded in plaster or run in shallow chases in masonry walls and covered with plaster. When so installed, if this cable is within 2 inches of the finished surface, it shall be protected against damage from nails by covering the chase with corrosion-resistant coated steel at least 1/16 inch thick and with a minimum width of 3/4 inch. This protective metal covering may be under the final surface finish (Figure 4-6).

NM and NMC cables shall be secured only with approved staples, straps, or similar fittings. These approved devices for the support of these cables are designed so as not to injure the cables.

These cables shall be secured within 12 inches of every cabinet, box or fitting and shall be secured at intervals of not over 4½ feet elsewhere. Of course, in wiring finished buildings, panels for prefabricated buildings and similar situations where the supports just mentioned are impractical, it is permitted to fish these cables between access points (Figure 4-7).

If it is desired to use NM cable in the air voids in masonry block walls (where, as previously stated, its use is prohibited), rigid conduit

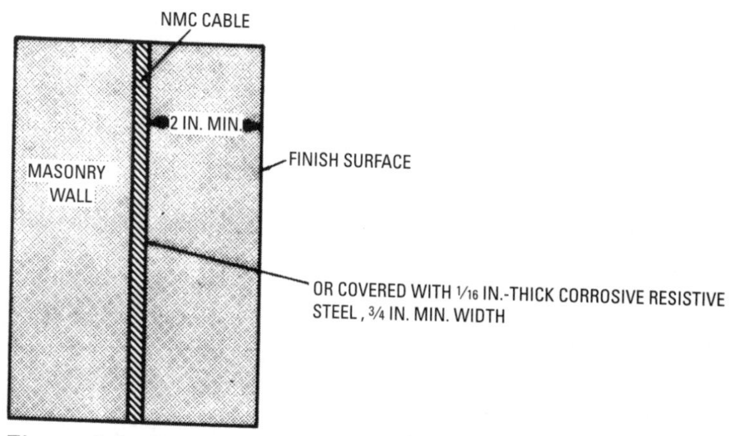

Figure 4-6 Proper installation of NMC cable in a plastered wall.

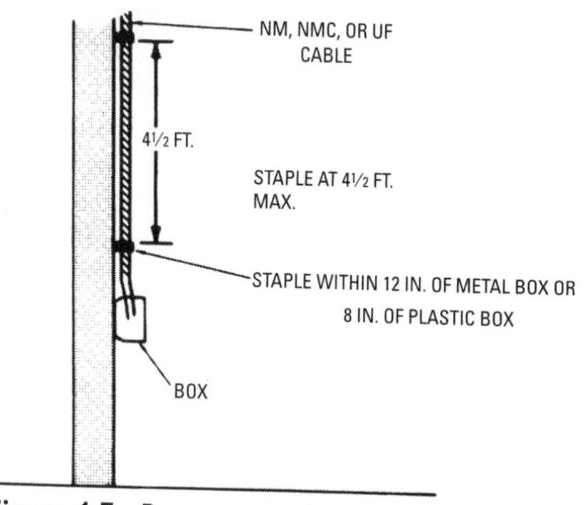

Figure 4-7 Proper method of securing NM, NMC, or UF cable.

or EMT may be installed in the voids first and then NM cable fished through these conduits or EMT.

In running NM or NMC cable in exposed work, the following requirements apply:

> *Following the surface. The cable shall closely follow the surface of the building finish or of running boards.*

Protection from physical damage. The cable shall be protected from physical damage where necessary by conduit, pipe, guard strips, or other means. Where passing through a floor the cable shall be enclosed in rigid metal conduit, intermediate metal conduit, electrical metallic tubing, or other metal pipe extending at least 6 inches (152 mm) above the floor.

Figure 4-8 illustrates the first of these *NEC* requirements, following the surface of the building. Figure 4-9 illustrates the second requirement, where protection from physical damage is required. Figure 4-10 illustrates where conduit or other protection is required when passing these cables through a floor.

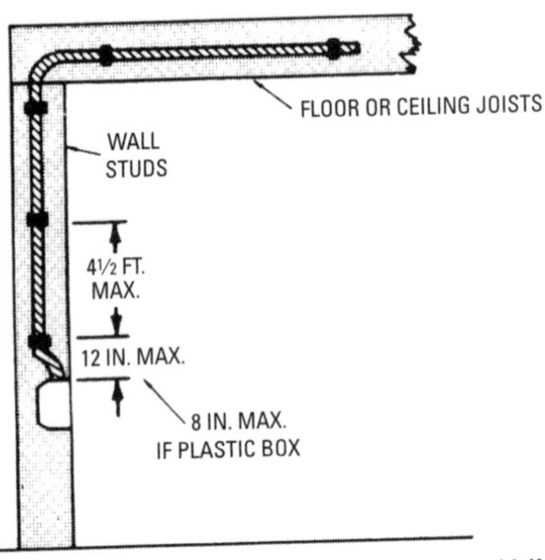

Figure 4-8 Proper installation of exposed NM or NMC cable following the surface of a building.

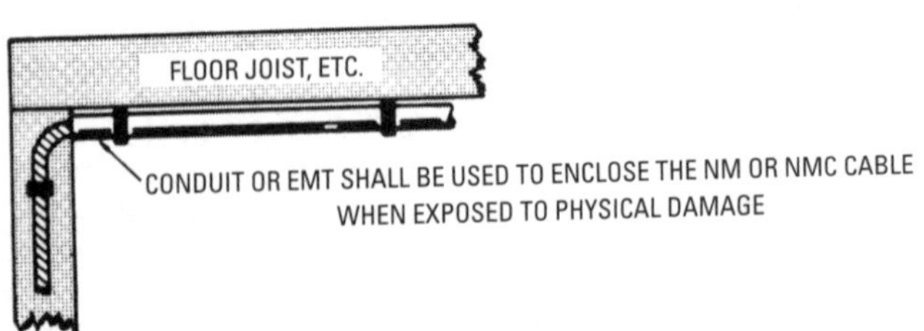

Figure 4-9 Installation method to protect NM or NMC cable from physical damage.

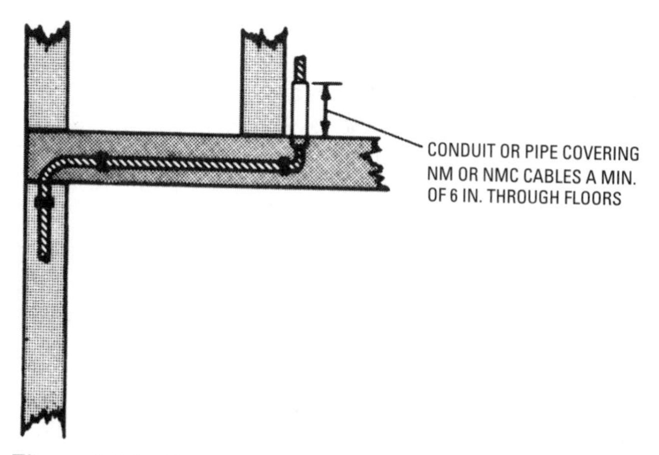

Figure 4-10 Protecting NM or NMC cable where it passes through a floor.

Now refer to *Section 334.15(C), In Unfinished Basements:*

> *Where the cable is run at angles with joists in unfinished basements, it shall be permissible to secure cables not smaller than two No. 6 or three No. 8 conductors directly to the lower edges of the joists; smaller cables shall either be run through bored holes in joists or on running boards.*

See Figures 4-10, 4-11, and 4-12.

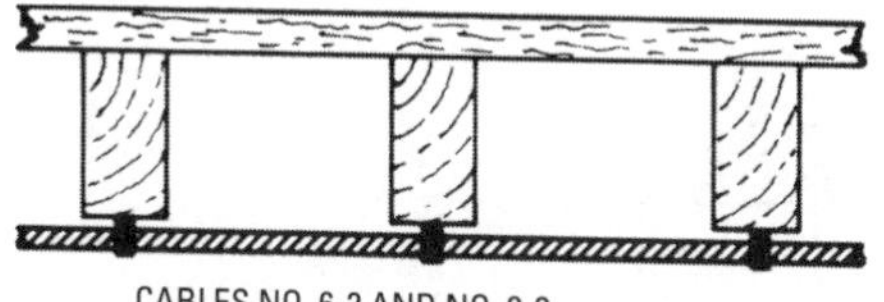

Figure 4-11 Large NM or NMC cable may be run across the lower edges of joists in unfinished basements.

Section 334.23 covers the requirements for installing NM cables in accessible attics. It does this by simply adopting the requirements of *Section 320.23* for Type AC cables (armored cables, discussed later in this chapter), which must be installed in accessible attics or roof spaces as specified in the following two paragraphs:

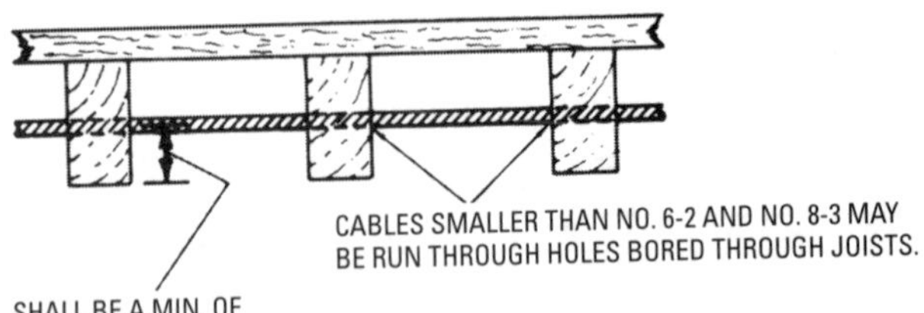

Figure 4-12 Small NM or NMC cable must be run through bored holes in the joists in unfinished basements or protected by running boards.

> *(A) Where Run Across the Top of Floor Joists. Where run across the top of floor joists, or within 7 feet (2.13 m) of floor or floor joists across the face of rafters or studding, in attics and roof spaces which are accessible, the cable shall be protected by substantial guard strips which are at least as high as the cable. Where this space is not accessible by permanent stairs or ladders, protection will only be required within 6 feet (1.83 m) of the nearest edge of the scuttle hole or attic entrance.*
>
> *(B) Cable Installed Parallel to Framing Members. Where cable is carried along the sides of rafters, studs, or floor joists, neither guard strips nor running boards shall be required.*

See Figures 4-13 and 4-14.

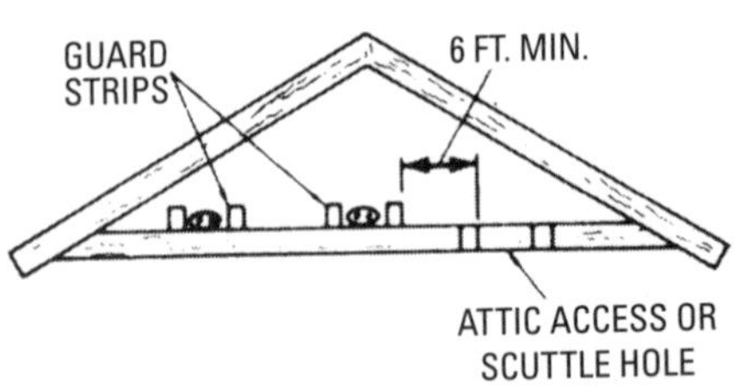

Figure 4-13 Guard strips are used to protect cable in accessible attics under certain conditions.

There are also requirements for cables installed through holes or notches in framing members:

> *Where exposed or concealed wiring conductors in insulating tubes or cables are installed through bored holes in studs, joists*

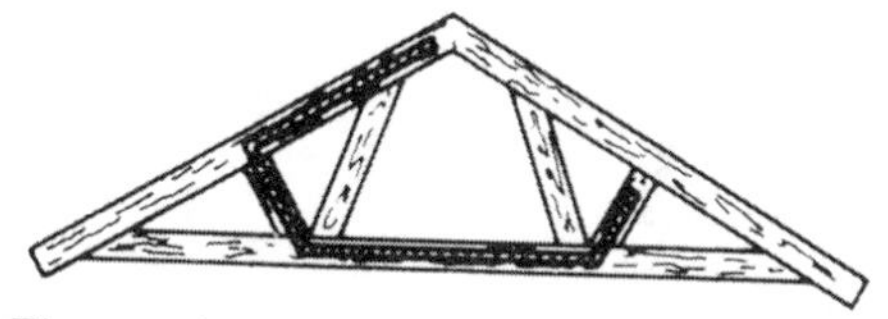

Figure 4-14 Cable run along the sides of rafters, studs, or joists in accessible attics needs no additional protection.

> *or similar wood members, holes shall be bored at the approximate centers of wood members, or at least 1¼ inches from the nearest edge where practical.*

The intent of the phrase "where practical" is in reference to a 2 × 4 stud (which in reality is only about 3⅝ inches but is considered to be 4 inches for all practical purposes) or similar building part. Bear in mind that, fundamentally, the 2-inch requirement is the prevailing factor.

> *Where there is no objection because of weakening the building structure, metal-clad or nonmetallic-sheathed cable, aluminum-sheathed cable and Type MI cable may be laid in notches in the studding or joists when the cable at those points is protected against the driving of nails into it by having the notch covered with a steel plate at least 1/16 inch thick before building finish is applied.* [See Figure 4-15.]

The primary purpose of all this is to protect the cables from being penetrated by nails or staples used in attaching lath, drywall, or paneling. Anyone who has ever had to troubleshoot a case caused by the penetration of a nail or staple into a cable will fully appreciate the intent of the *NEC* on this point.

In the installation of NM and NMC cables, *Section 334.24, Bending Radius,* shall be adhered to. This section states:

> *Bends in cable shall be so made, and other handling shall be such, that the protective coverings of the cable will not be injured, and no bend shall have a radius less than five times the diameter of the cable.*

NM and NMC cables can be damaged due to sharp bending, or damage to the other covering caused by pulling these cables through bored holes. Problems may also occur by driving too severely the staples holding the cables.

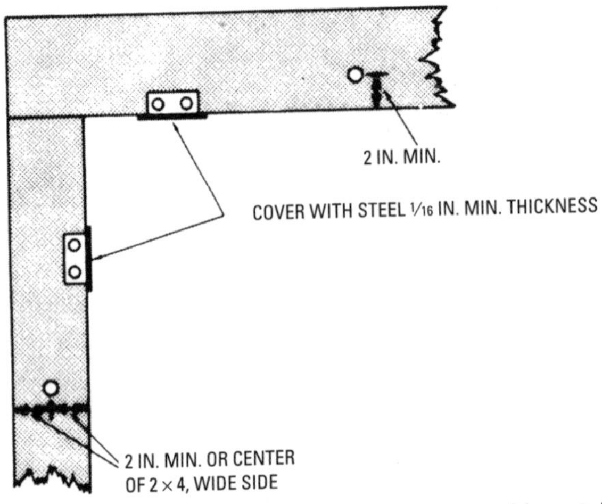

Figure 4-15 Steel plates used to protect cable notched into studs or joists.

Underground Feeder and Branch-Circuit Cable (Type UF)

UF cables are by far the most common residential wiring method where wiring must be installed underground. UF cables are suitable for smaller circuits and in most locations. But although UF cable is fine for burying, it is considered good trade practice to protect these cables with some sort of conduit or raceway as they emerge from the ground.

Use and Installation of UF Cable

Article 340 of the *NEC* covers Type UF cable. This is an underground feeder or branch-circuit cable of an approved type that comes in sizes No. 14 to 4/0 AWG, inclusive. The covering on the cable is:

1. Flame retardant
2. Moisture resistant
3. Fungus resistant
4. Corrosive resistant
5. Suitable for direct burial in the earth

Section 340.10 details the areas where UF cables may be installed. We need not go into the entire section, but we will cover that portion we consider applicable to residential wiring.

Fundamentally, Type UF cable may be buried directly in the earth and may be made up of a single conductor or multiple conductors in one sheath. We will first discuss burying this cable in the earth.

Underground burial requirements are found in *Table 300.5*, which mandates a minimum depth of 24 inches to be maintained for conductors and cables buried directly in the earth. According to *Table 300.5*, this depth may be reduced to 18 inches, provided that supplemental protective covering such as a 2-inch concrete pad, metal raceway, pipe, or other suitable protection is used (Figure 4-16).

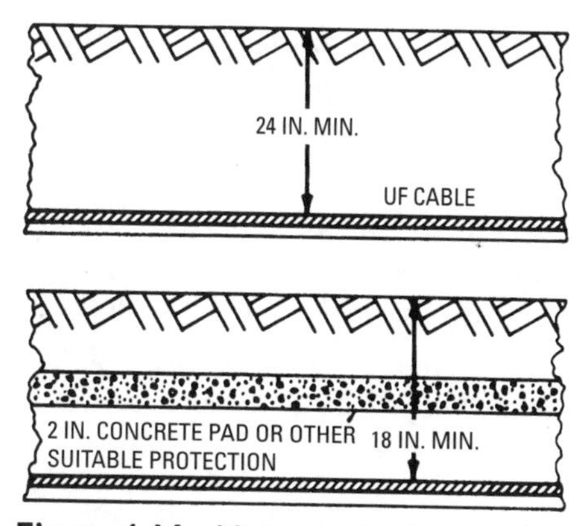

Figure 4-16 Minimum depth at which UF cable must be buried.

Now let us analyze this portion of the *NEC*. We find the minimum depth of burial is 24 inches. It will be found that if this cable is buried in a rock-type soil, the inspector will, without doubt, require the cable to have a fine sand bed placed in the bottom of the ditch, with a layer of fine sand required as the first cover over this cable. This point is especially important in climates where the ground is subject to freezing at this depth or deeper, the intent being to protect the cable insulation from damage due to the rocks in the soil (Figure 4-17).

This 24-inch depth may be reduced to 18 inches if there is a 2-inch concrete pad, metal raceway, or other means of protecting the cable as called for. This cable must enter the ground at some point and must also emerge at some point. At these points the UF cable

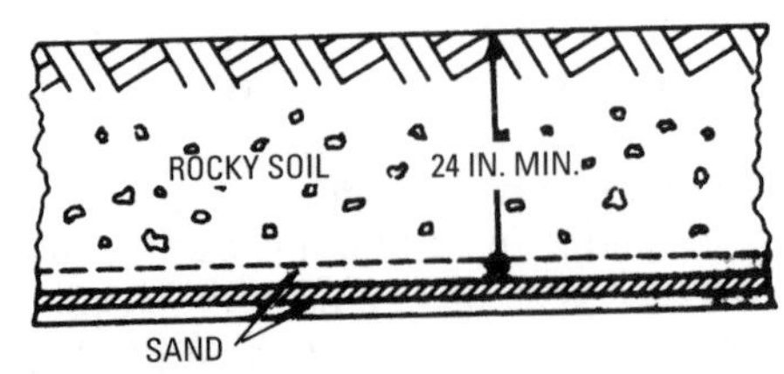

Figure 4-17 Method of burying UF cable in rocky soil.

must be sleeved in metal raceways and proper protection given so that frost heave will not cause cutting of the cable against the metal raceway. This can, of course, be done by using plastic insulating bushings at these points.

Section 340.12 says that this cable shall not be used when *embedded in poured cement, concrete, or aggregate, except where embedded in plaster as nonheating leads as provided in Section 424.43.*

> *300.5(C). Underground Cables under Buildings. Underground cable installed under a building shall be in a raceway that is extended beyond the outside walls of the building.*
>
> *300.5(D). Protection From Damage. Direct-buried conductors shall be protected against physical damage in accordance with (1) through (5) [of Section 300.5(D)].*

The following methods are available for protecting UF cable:

1. In duct
2. In rigid metal conduit or electrical metallic tubing made of a material suitable for the condition, or provided with corrosion protection suitable for the condition
3. In rigid nonmetallic conduit if installed in accordance with *Article 352*
4. In cable of one or more conductors approved for direct burial in the earth
5. By other approved means

We may ignore item 4 if we are talking about underground cable run under a building, because we are told in *Section 300.5* that it shall be in a raceway.

In other words, any UF cable buried under a building must be installed so that it can be readily replaceable. You will also find that inspection authorities will, for the most part, consider concrete patios and the like to be a part of the building. To clarify this, they would consider it impractical to put a portion of the UF cable under the building proper in a raceway and not have it in a raceway under a concrete patio. There would be no way to replace the UF cable under the patio, should it become necessary to do so, without tunneling. So it makes good sense to require the raceway to extend out to the point where the cable may be reached by digging in exposed earth (Figure 4-18).

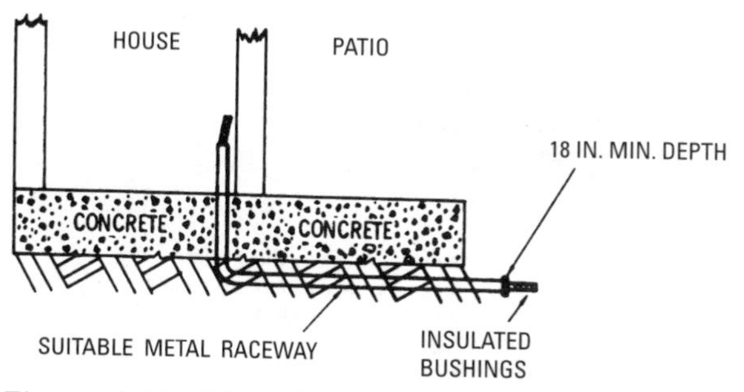

Figure 4-18 UF cable buried under a building or other inaccessible locations must be installed so as to be readily replaceable.

The installation of UF cable will be done in exactly the same way as the installation of NM cables. Note that UF (as well as NMC cable) may be run in the voids of block walls exposed to dampness.

Metal-Clad Cable (Type AC)

Type AC cable shares the advantages of Type NM cable in that it can be installed quickly and easily by hand. No special benders or measurements are required. But AC cable also has armor, making it tougher and more durable than NM cable. A stray nail, driven into a wall while hanging a picture, can pierce and short out an NM cable. The same nail would not likely penetrate the armored cable.

Use and Installation of AC Cable

Article 320 covers Armored, or Type AC, cable. What we are interested in here is in regard to residential wiring with Type AC cable.

Type AC cable is a fabricated assembly of insulated conductors in a metallic enclosure. *Section 320.100* defines AC cable (illustrated here in Figure 4-19):

> *320.100. Construction. Type AC cable shall have an armor of flexible metal tape, and shall have an internal bonding strip of copper or aluminum, in intimate contact with the armor for its entire length.*

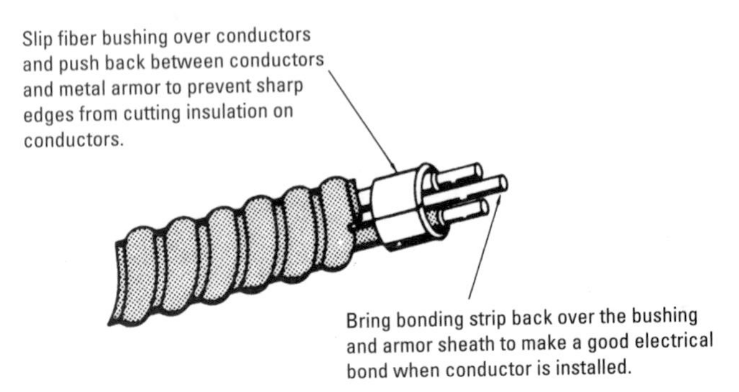

Figure 4-19 Type AC metal-clad cable.

Type AC cable is widely used in residential wiring. You will commonly hear it referred to as BX—an old reference to the first manufacturer of this cable type, the Bronx Cable Company.

According to *Section 320.10*, metal-clad cable of the AC type may be used in dry locations and may be embedded in plaster finish on brick or other masonry, except in damp or wet locations. This cable may be run or fished in the air voids of masonry block or tile walls, except where such walls are exposed or subject to excessive moisture or dampness or are below grade line. Most inspection authorities would not look favorably on the installation of Type AC cables in chases of outside brick walls exposed to the dampness of weather conditions.

The requirements for installation of Type AC cable are very similar to those for Type NM, covered earlier in this chapter. Bear in mind, however, that at each termination of AC cable, a fiber bushing made for this purpose shall be inserted between the armored sheath and the conductors. It should also be noted that either the grounding strip, as referred to in the construction of Type AC cable, shall be bent back over the fiber bushing and in intimate contact with the external portion of the armor and the cable clamp so installed that this grounding strip is made electrically secure by the clamp, or else this grounding strip shall be installed in the same way as the equipment grounding conductor in NM cable (Figure 4-20).

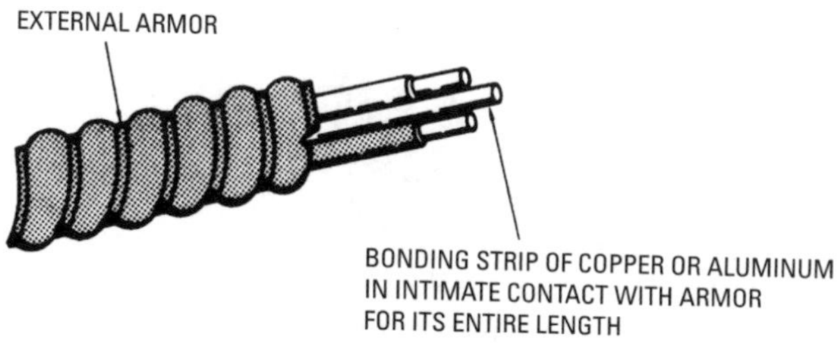

Figure 4-20 Proper method of preparing AC cable for installation.

The supports for Type AC cable are the same as for NM cable, namely, at intervals not exceeding 4½ feet and within 12 inches of every outlet box or fitting. Of course, we cannot staple where Type AC cable is fished in walls. Also, lengths not exceeding 24 inches may be used at terminals if required. The boxes and fitting shall be approved for Type AC cable. All other conditions of installation shall be the same as for Types NM, NMC, and UF cables for the interior wiring of a residence, unless there are exceptions, which were noted in the first part of this chapter.

Service-Entrance Cable (Types SE and USE)

Type SE cable makes for a less expensive installation than conduit with conductors. But it is also far less durable and more susceptible to damage. Because of this, the installation of services in raceway is much preferred in the electrical trade. SE cable is allowed, but for relatively small residential services only. SE cable is generally used where cost is an overriding concern.

> *338.2. Definition. Service-entrance cable is a conductor assembly provided with or without an overall covering, primarily used for services and of the following types:* [See Figure 4-21.]
>
> *Type SE. Service Entrance cable having a flame-retardant, moisture-resistant covering.*
>
> *Type USE. Service Entrance cable, identified for underground use, having a moisture-resistant covering, but not required to have a flame-retardant covering.*

Type SE cable is so designated because it is often used for service-entrance conductors. Refer back to Chapter 2, which

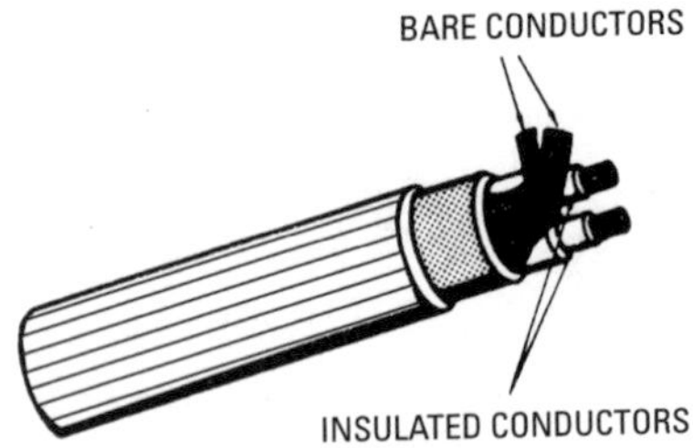

Figure 4-21 SE cable with a bare neutral conductor.

covers in details the installation of service entrances and service-entrance conductors.

Use and Installation of SE and USE Cables

Obviously Type SE and USE cables may be used for service entrances, but *Section 338.10(B)* also allows them to be used as branch-circuit cables. As branch circuits, SE cables are commonly used to provide 40- and 50-amp circuits to electric ranges and to HVAC units. When used this way, you will sometimes hear SE referred to as *range cable.*

> *338.10(B)(1) Grounded Conductor Insulated. Type SE service-entrance cables shall be permitted in interior wiring systems where all of the circuit conductors of the cable are of the rubber-covered or thermoplastic type.*

Take note of the phrase "where all of the circuit conductors of the cable are of the rubber-covered or thermoplastic type." Figure 4-22 shows such a cable. The cable contains three insulated conductors to be used as circuit conductors, namely for wiring 120/240-volt systems, which are the type usually used in residential wiring. Two of the insulated conductors are phase or current-carrying conductors, and the third insulated conductor is the neutral. There should be one stranded bare conductor, which is used as an equipment-grounding conductor. In a majority of uses for branch circuits and feeders, the equipment-grounding conductor will be required. Note that *Section 338.10(B)(2)* also states that the foregoing provisions do not intend to deny the use of service-entrance cable for interior use when the fully insulated conductors are used for circuit wiring and the uninsulated conductor is used for equipment grounding purposes.

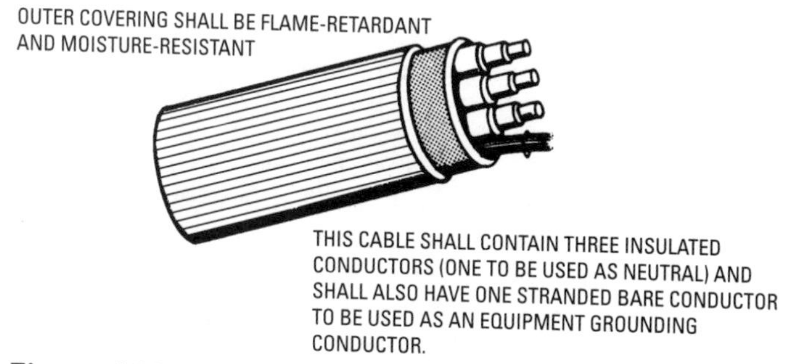

Figure 4-22 SE cable with insulated conductors and an equipment-grounding conductor.

An explanation of this statement is as follows, referring to the Type SE cable illustrated in Figure 4-21. For instance, a 240-volt branch circuit or feeder might be used to supply 240-volt loads that do not require a neutral conductor. These could be electrical heating panels or electrical heating equipment. This would require the use of the two insulated conductors for the 240-volt supply and the bare conductor would be used as an equipment-grounding conductor. You are cautioned to observe this information very closely, because for some reason the use of SE cables seems to be confusing at times.

Caution should be exercised when using the four-conductor type for a service entrance. In this type of cable, there is an insulated neutral plus two phase conductors and a bare grounding conductor. The common practice is to connect both the bare equipment-grounding conductor and the insulated neutral conductor to the meter-housing neutral terminal and at the neutral bus in the service equipment. This is a violation of *Section 310.4, Conductors in Parallel,* which lists the criteria for paralleling conductors. This type of cable does not have a neutral conductor and an equipment-grounding conductor to meet the criteria of *Section 310.4*. Therefore, the bare equipment-grounding conductor should be eliminated when this type of cable is used for service-entrance conductors.

Section 338.10(B)(2) states:

> *Type SE service-entrance cables shall be permitted for use where the insulated conductors are used for circuit wiring and the uninsulated conductor is used only for equipment grounding purposes.*

We find some of the same material here that was previously explained in this chapter. A cable with insulated conductors and one bare conductor, as illustrated in Figure 4-21, may be used as a branch circuit to supply a range, wall-mounted oven, counter-mounted cooking unit, or clothes dryer. A three-conductor cable (SE), with one conductor bare, may be used to supply ranges and the like only when this branch circuit originates in service-equipment panels (Figure 4-23). In case this is confusing, the reasoning is as follows: In the service equipment, the neutral bus is to be grounded to the panel enclosure. Therefore, when the bare neutral of the Type SE cable under discussion is used, hazardous conditions are created, because if the bare neutral happens to touch the enclosure, they are electrically connected together in the service equipment. When supplying ranges and similar appliances from a feeder panel, however, an entirely different condition exists. The neutral bus in the feeder panel is isolated from the enclosure, as shown in Figures 4-24 and 4-25. In Figure 4-24, the service equipment and a feeder panel are connected by a four-conductor SE cable having three insulated conductors and one bare equipment-grounding conductor. In the service equipment the neutral bus is bonded to the service-equipment enclosure, and, in addition, the insulated neutral and the bare equipment grounding conductor of the SE cable are both tied to the neutral bus. In the feeder panel the neutral bus is isolated, and the insulated neutral goes to this neutral bus. Notice that the grounding bus in the feeder panel enclosure is tied to the enclosure, and it is to this bus, not the neutral bus, that the bare equipment-grounding conductor is connected. In case someone thinks this is paralleling the neutral and the equipment-grounding conductor, it should be stated that this is not true. Although these two conductors are tied together in the service-equipment enclosure, they are not tied together in the feeder panel, so they are not paralleled.

In Figure 4-25, either a four-conductor cable or a three-conductor cable with all three conductors insulated is used as the branch circuit to a range or dryer. What do we have? With a four-conductor cable the neutral and the equipment ground conductor are separated; but the neutral is an insulated conductor, so that if it comes in contact with the enclosure (which it most likely will), the purpose of isolating

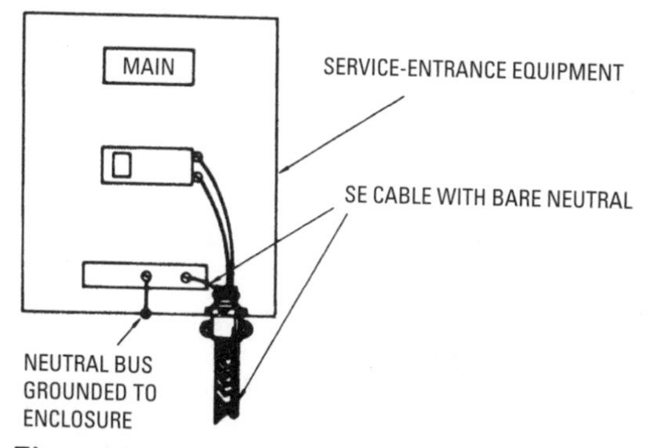

Figure 4-23 SE cable with a bare neutral conductor may be used as a branch circuit to supply ranges, dryers, and similar appliances only when it originates in service-equipment panels.

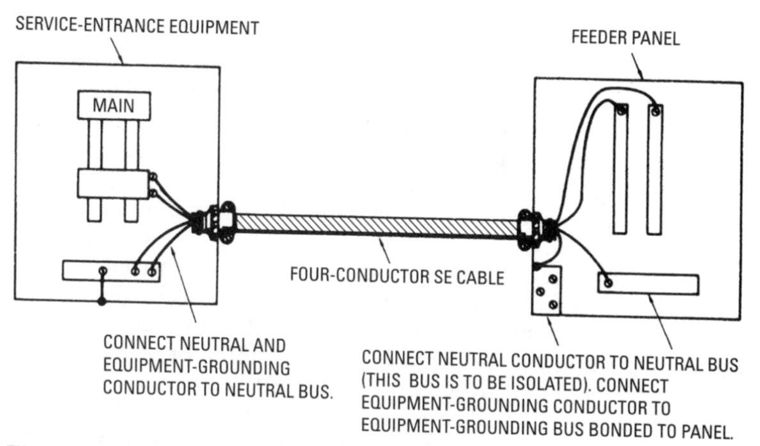

Figure 4-24 SE cable with three insulated conductors and a bare equipment-grounding conductor must be used when supplying a feeder panel.

the neutral bus in a feeder panel is not defeated. If a three-conductor insulated SE cable is used, the *Code* permits us to ground the frame of the range or dryer to the neutral in residences. *Caution:* If a four-conductor cable with one conductor bare is installed, use the bare conductor for grounding the frame of the range or dryer, but do not ground the neutral to the frame in this case. To do so would defeat the purpose.

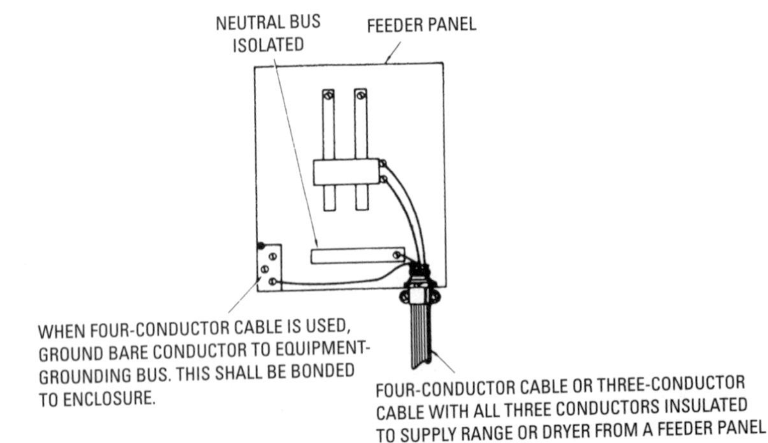

Figure 4-25 Branch-circuit connections from a feeder panel.

Section 338.10(B)(4) allows SE cable with a bare neutral, as illustrated in Figure 4-21, to be used as a feeder. This is not recommended. If you are going to use any type of cable as a feeder, use an insulated neutral and a bare equipment-grounding conductor. The cost difference between the cables is minimal, and the use of bare neutrals is fraught with difficulty. A detailed explanation follows:

In a residence, we might have a detached garage. This could be supplied from the panel in the house by means of a feeder circuit using SE cable with two insulated conductors and a bare neutral but with no equipment-grounding conductor. This is confusing because, in dealing with feeders, we normally think of an equipment-grounding conductor having to be supplied with the feeder. However, where more than one building (such as a garage) is supplied by the same service, the grounded circuit (neutral) conductor of the wiring system of any building utilizing one branch circuit supplied from the service *may* be connected to a grounding electrode at this building. In the case of any building housing equipment required to be grounded or utilizing two or more branch circuits supplied from such service, and in the case of a building housing livestock, the grounded circuit conductor *shall* be so connected. Thus, even though we have a feeder circuit, the other building is treated as a service for wiring purposes, and a ground electrode is provided at the second building.

A word of caution is given at this point on checking the markings on SE cable. The outer surface of SE cable shall be marked at intervals not exceeding 24 inches. This marking shall designate the type of cable, the size of the conductors, and the type of insulation that is used on these conductors. This is important when using *Table 310.16* to determine the allowable ampacities of conductors. In this table we find that No. 8 aluminum conductors with 60°C (140°F) insulation have ampacity of 30, while those with 75°C (167°F) insulation have ampacity of 40, and so forth, so the overcurrent protection will have to fit this table.

When installing SE and USE cable, USE cable shall be treated in the same manner as UF cable. SE cable comes under the provisions of not only *Article 338* but also *Article 300* and *Article 340*. Most of these requirements were covered earlier in this chapter, so they will not be repeated at this point.

Electrical Metallic Tubing

Electrical metallic tubing, commonly known as EMT or thin-wall conduit, is far less used in residential wiring than are NM and AC cables. This is because its installation requires more labor and more specialized skills. EMT must be bent with a special tool and pieced together properly. EMT is, however, an excellent wiring method. It is durable, protects conductors quite well from damage, and allows future changes to a wiring system by merely pulling old conductors out of the raceway and pulling new conductors in.

EMT is considered an excellent wiring method for house wiring, but it is actually used primarily in those municipalities where it is required and for high-end homes.

Use and Installation of EMT

Electrical metallic tubing appears in *Article 358* of the *NEC*. EMT is probably the third most widely used wiring method for residential wiring, behind only nonmetallic-sheathed cable and, possibly, Type AC cable. EMT is not technically conduit, but it is treated similarly by the *NEC*. It shares many of the same requirements that apply to rigid, intermediate, and other metallic conduit systems.

Section 358.10 outlines the uses of EMT. It may be used in almost any common location, with the caveat that corrosion protect must be used if EMT is directly buried or in wet locations. You will also find that most inspection authorities will never permit EMT in contact with the earth, and under *Section 90.4* it becomes their right to judge the conditions. You will find some inspection authorities never permit EMT in concrete slabs, while others will permit such an installation provided that the concrete does not have an additive such as calcium chloride.

EMT is manufactured in sizes from ½-inch through 4-inch electrical trade size. There is an exception for tubing containing motor leads, but this is extremely rare, and would never show up in a house anyway.

The numbers of conductors allowed in EMT are covered in *Section 358.22*. This section refers us to *Table 1, Chapter* 9 of the *NEC*, reproduced here in part as *Table 4-2*. See also *Table 4* of *Chapter 9* of the *NEC*. For the maximum wire fills of EMT, refer to *Table C1* of *Annex* C of the *NEC*, reproduced here as *Table 4-3*. Maximum percentages of the cross-sectional area of EMT or other conduit are given in *Table 4-4*.

Derating

When more than three current-carrying conductors are contained in a conduit or raceway (such as EMT), their current-carrying capacity (ampacity) must be derated. You can avoid this issue entirely by not installing more than three current-carrying conductors in any piece of EMT, which is generally a good practice anyway.

Table 4-5, taken from *Table 310.15(B)(2)* of the *NEC*, details the derating amounts.

To understand this further, follow this example: Turn to *Table 310.16* and use the 75°C (167°F) column for the proper conductors. Using No. 14 AWG, we find that it has an allowable ampacity of 15 amperes. Using *Table 4-5*, we find that if we do not have over three No. 14s in a raceway, we may use 15-ampere overcurrent devices on this branch circuit, but if we have four No. 14s in a raceway, we have to derate to 80 percent of the 15 amperes, which would be 12 amperes. Since this is an impractical combination (no 12-amp circuit breakers are made), we changed to No. 12 AWG, which, according to *Table 310.16*, gives us an allowable ampacity of 20 amperes. Since we are using four conductors in a raceway, *Table 4-5* tells us we must derate to 80 percent of the 20 amperes, which equals 16 amperes. Therefore, we will be required to use No. 12 copper wire on this 15-ampere branch circuit. The same type of calculations would be used for aluminum conductors.

Your attention is called to the note at the head of both *Tables 310.16* and *310.18: Not More than Three Conductors in Raceway or Cable or Earth (Directly Buried) Based on Ambient Temperature of 30°C [86°F]*.

"Ambient temperature" refers to the temperature of the surrounding area in which the conductors and raceways are installed.

Table 4-2 Maximum Number of Conductors in Trade-Sizes of Conduit or Tubing (Based on *Table 1, Chapter 9*, in the *NEC*)

Conduit Trade Size (Inches)											
Type Letters	**Conductor Size AWG, kcmil**	**1/2**	**3/4**	**1**	**1 1/4**	**1 1/2**	**2**	**2 1/2**	**3**	**3 1/2**	**4**
	14	12	22	35	61	84	138				
THHN, THWN, THWN-2	12	9	16	26	45	61	101	176			
	10	5	10	16	28	38	63	111	167		
	8	3	6	9	16	22	36	64	96	106	136
	6	2	4	7	12	16	26	46	69	76	98
	4	1	2	4	7	10	16	28	43	47	60
	3	1	1	3	6	8	13	24	36	39	51
	2	1	1	3	5	7	11	20	30	33	43
	1		1	1	4	5	8	15	22	25	32
	0		1	1	3	4	7	10	19	21	27
	00		1	1	2	3	6	8	16	17	22
	000		1	1	1	3	5	7	13	14	18
	0000		1	1	1	2	4	6	11	12	15
	250			1	1	1	3	5	9	10	12
	300			1	1	1	3	4	7	8	11
	350			1	1	1	2	3	6	7	9
	400				1	1	1	3	6	6	8
	500				1	1	1	2	5	5	7
	600				1	1	1	1	3	4	5
	700					1	1	1	3	4	5
	750					1	1	1	3	3	4

Table 4-3 Maximum Number of Conductors in EMT (*Table C.1*, Based on *Table 1*, *Chapter 9*, in the *NEC*)

Conduit Trade Size (Inches)											
Type Letters	**Conductor Size AWG, kcmil**	**1/2**	**3/4**	**1**	**1 1/4**	**1 1/2**	**2**	**2 1/2**	**3**	**3 1/2**	**4**
TW, THHW, THW, THW-2	14	8	15	25	43	58	96	168	254	332	
	12	6	11	19	33	45	74	129	195	255	
	10	5	8	14	24	33	55	96	145	190	
	8	2	5	8	13	18	30	53	81	105	
	6	1	3	4	8	11	18	32	48	63	81
	4	1	1	3	6	8	13	24	36	47	60
	3	1	1	3	5	7	12	20	31	40	52
	2	1	1	2	4	6	10	17	26	34	44
	1		1	1	3	4	7	12	18	24	31
	0		1	1	2	3	6	10	16	20	26
	00		1	1	1	3	5	9	13	17	22
	000		1	1	1	2	4	7	11	15	19
	0000			1	1	1	3	6	9	12	16
	250			1	1	1	3	5	7	10	13
	300			1	1	1	2	4	6	8	11
	350				1	1	1	4	6	7	10
	400				1	1	1	3	5	7	9
	500				1	1	1	3	4	6	7
	600					1	1	2	3	4	6
	700					1	1	1	3	4	5
	750					1	1	1	3	4	5

Table 4-4 Percent of Cross Section of Conduit and Tubing for Conductors

Number of Conductors	1	2	3	4	Over 4
All conductor types except lead-covered (new or rewiring)	53	31	40	40	40
Lead-covered conductors	55	30	40	38	35

Table 4-5 Derating Conductors in EMT

Number of Conductors	Percent of Values in Tables 310.16 and 310.318 as Adjusted for Ambient Temperature if Necessary
4 through 6	80
7 through 24	70
25 through 42	60
43 and above	50

Tables 310.16 through 310.19 each have derating tables at the bottom of the ampacity tables that must be used for high-temperature areas.

EMT is not to be coupled together nor connected to boxes or fitting by means of threads made in the wall of the EMT. None but approved fittings are to be used for the purpose. In using EMT connectors or couplings, be certain that they have the UL letters on them, or at least on the packaging in which they come.

Figure 4-26 shows a pressure-type coupling. Each end has a nut that, when screwed down on the coupling threads, puts pressure against a spring device within the nut so as to make a good electrical contact.

Figure 4-27 shows a setscrew-type connector. Couplings of this type are also available. Note the setscrew that is tightened down against the EMT inserted into the coupling or connector, thus making a good electrical contact.

EMT couplings and connectors shall be made tight for good electrical continuity. When buried in masonry or concrete, they shall be of the concrete-tight type; and when used in wet locations, they shall be of the raintight type. A description of the purposes for which, and locations where, they may be used will be found on the cartons in which they are packed.

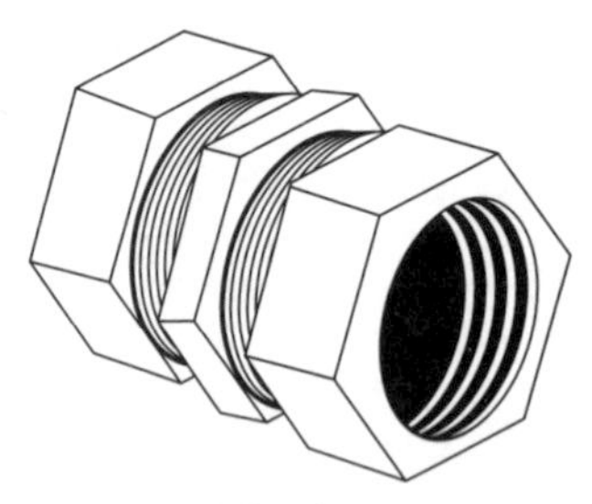

Figure 4-26 A pressure-type EMT coupling.

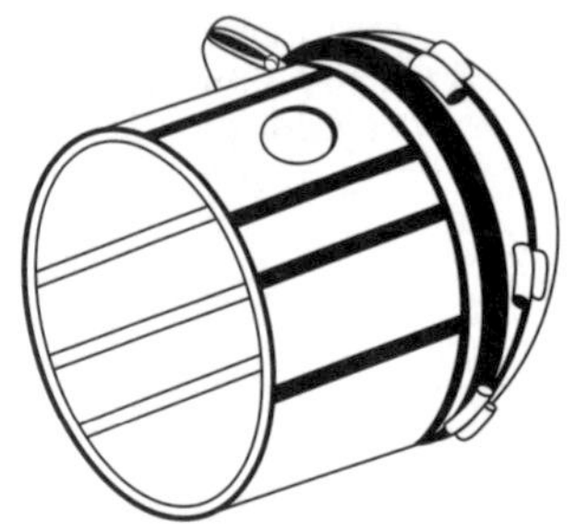

Figure 4-27 A setscrew-type EMT connector.

> *358.26. Bends—Number in One Run. A run of electrical metallic tubing between outlet and outlet, between fitting and fitting, or between outlet and fitting, shall not contain more than the equivalent of four quarter bends (360° total), including those bends located immediately at the outlet or fitting.*

Figure 4-28 shows two boxes connected by a raceway with four 90° bends. This makes a total of 360°, which is the maximum permitted by the *Code*. Figure 4-29 also shows a total of 360° total bends between the boxes. This, however, differs from Figure 4-28 in that two 45° offset bends were made at each box to bring the EMT from the knockout in the box to where it could be strapped to a wall. We also find two 90° bends, so we have used the allowable total number of bends permitted between boxes. The purpose of this 360° total is that it was apparently felt that more bends than this total amount would cause difficulty in pulling conductors into the raceway and possibly cause damage to the insulation.

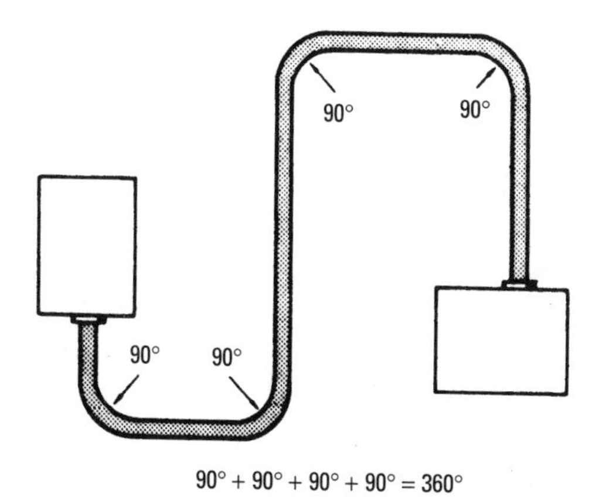

Figure 4-28 The maximum amount of bends in EMT must not exceed 360° between any two boxes or pull points.

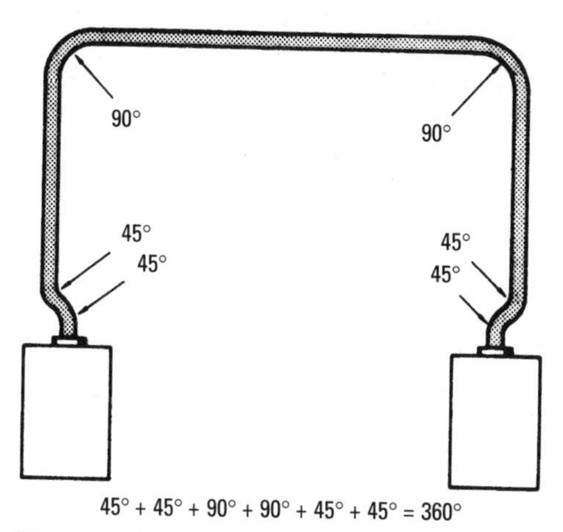

Figure 4-29 Another example of the maximum amount of bends permitted in EMT between junctions.

> *358.24. Bends—How Made. Bends shall be made so that the tubing is not damaged and the internal diameter of the tubing is not substantially reduced. The radius of the curve of any field bend to the center line of the conduit shall not be less than shown in Table 344.24 for one-shot and full shoe benders.* [See Figure 4-30.]

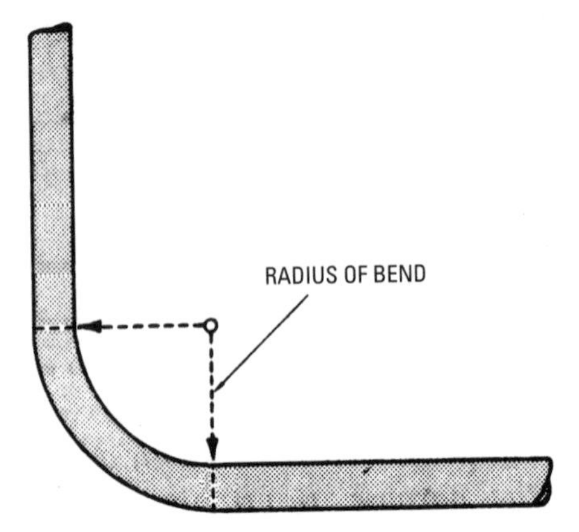

Figure 4-30 The radius of bend in EMT.

Bending EMT (and other conduits) must be performed with benders made for the purpose. Otherwise, the tubing or conduit will likely be malformed, making the installation of conductors difficult and possibly leading to abrasions on the conductors, and from there to short circuits (Figure 4-31). An insufficient radius on the bend also makes the installation of conductors overly difficult, requiring excessive pulling forces.

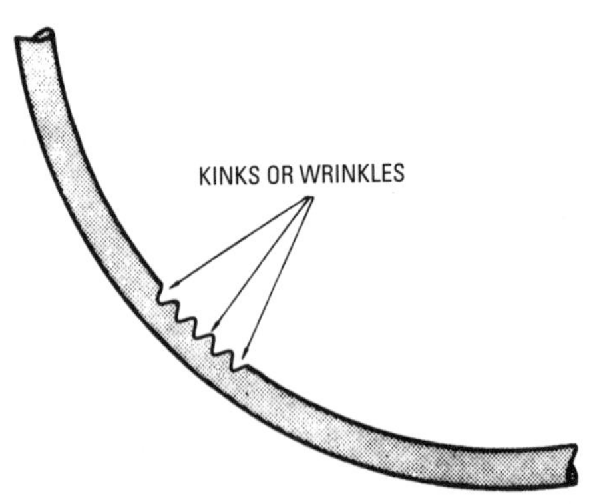

Figure 4-31 Improper bending may produce wrinkles or kinks in EMT.

> *358.28(A). Reaming. All cut ends of electrical metallic tubing shall be reamed to remove rough edges.*

In Figure 4-32, note that when EMT is cut with a wheel-type pipe cutter, the tubing tends to roll in. All of this must be cut away by means of a reamer. EMT cut with a hacksaw does not have this rolled-in burr, but care must be taken to make square cuts and then ream to eliminate any burrs and give a beveled edge. It is impossible for an inspector to see every cut on a wiring job, but be assured that if this reaming is not done properly, breakdown or damage to the insulation will occur.

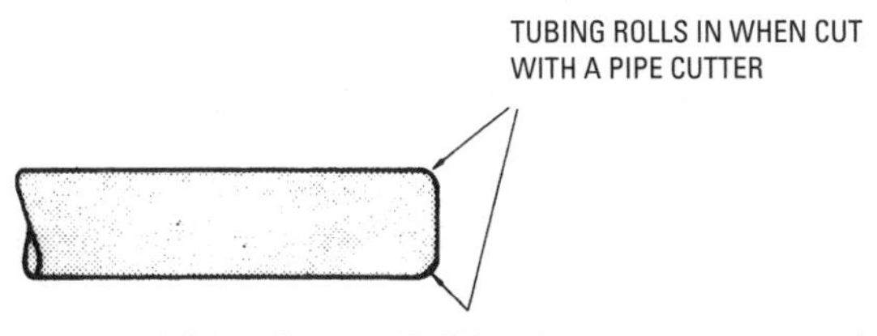

Figure 4-32 Cutting EMT with a pipe cutter produces a rolled-in burr.

> *358.80. Securing and Supporting. Electrical metallic tubing shall be installed as a complete system as provided in Article 300 and shall be securely fastened in place at least every 10 feet (3.05 m) and within 3 feet (914 mm) of each outlet box, junction box, cabinet or fitting.* [See Figure 4-33.]

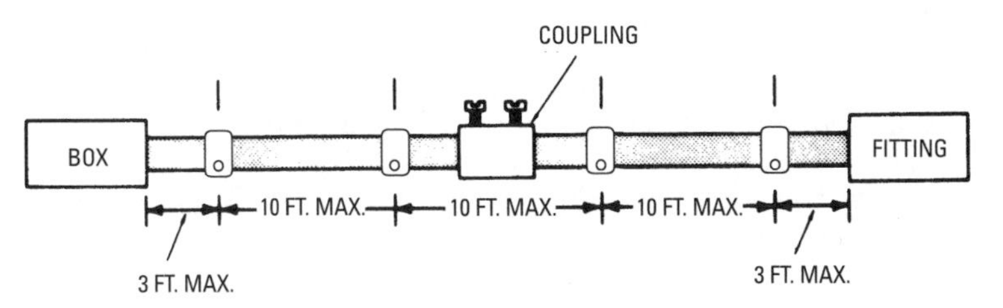

Figure 4-33 EMT must be supported at specified maximum intervals.

Rigid Metal Conduit

Rigid metal conduit (commonly called *RMC* or *GRC – galvanized rigid conduit*) is seldom used in house wiring except for the service entrance. Rigid metal conduit has the best protective and durability characteristics of any wiring method, but it is expensive and requires more specialized labor to install.

Use and Installation of Rigid Metal Conduit

Rigid metal conduit is covered in *Article 344* of the *NEC*. Much of the preceding section and *Article 358* are also quite applicable to rigid metal conduit.

Rigid metal conduit may be used in all atmospheric conditions and occupancies except that ferrous metal raceways and fittings protected only by enamel may be used only indoors and in occupancies not subject to a corrosive influence. We must also bear in mind galvanic action (electrolysis) and its effect on dissimilar metals.

> *344.10(B). RMC elbows, couplings, and fittings shall be permitted to be installed in concrete, in direct contact with the earth, or in areas subject to severe corrosive influences where protected by corrosion protection judged suitable for the condition.*

This paragraph automatically places the question of conditions and locations where rigid metal conduit may or may not be installed in earth or in concrete up to the authority having jurisdiction.

There is factory-coated conduit available, and so are materials that may be sprayed or painted on the conduit and will be satisfactory for approval. Check with the inspection authority and see what it will require.

You will find that, to a large degree, inspection authorities will never permit rigid metal conduit in contact with the earth and, under *Section 90.4*, it becomes their right to judge the conditions. However, you will find that most inspection authorities will permit rigid metal conduit in concrete without additional protection from corrosion unless the concrete has an additive such as calcium chloride. An inspector recently observed rigid galvanized conduit that had been in a concrete slab for several years. Calcium chloride had originally been added to the concrete to prevent freezing. When this concrete was broken up, it was found that the conduit was very badly eaten away. Practically no inspector will permit aluminum conduit in the earth or in concrete, as it seems to be especially vulnerable to corrosion in these areas. In deference to aluminum conduit, however, there are corrosive influences (but not in residential wiring locations) that do not affect aluminum nearly as much as they affect rigid galvanized conduit.

These are the most basic requirements pertaining to RMC:

- *344.20(A). Minimum Size.* No conduit smaller than 1/2 inch, electrical trade size, shall be used.

- *344.22. Number of Conductors in Conduit.* Refer to the coverage given to EMT. Everything stated there, including references to the tables, are all applicable to rigid metal conduit.
- *344.28. Reaming and Threading.* The requirement is the same as for the reaming of EMT. Conduit threaded in the field is threaded with a standard conduit-cutting die with a 3/4-inch taper per foot.
- *344.46. Bushings. Where a conduit enters a box or other fitting, a bushing shall be provided to protect the wire from abrasion unless the design of the box or fitting is such as to afford equivalent protection.*

(FPN): See Section 300.4(F) for the protection of No. 4 and larger conductors at bushings.

Where ungrounded conductors of No. 4 or larger enter a raceway or cabinet, pull box, junction box, or auxiliary gutter, the conductors shall be protected by a substantial bushing providing a smoothly rounded insulating surface, unless the conductors are separated from the raceway fitting by substantial insulating material securely fastened in place.

Figure 4-34 shows insulated bushings and locknuts to be used for this purpose. Figure 4-35 illustrates a fiber insert to be used with standard metal bushings. Where bonding-type bushings are to be used, they can be metal with an insert, or they may be metal with a plastic throat.

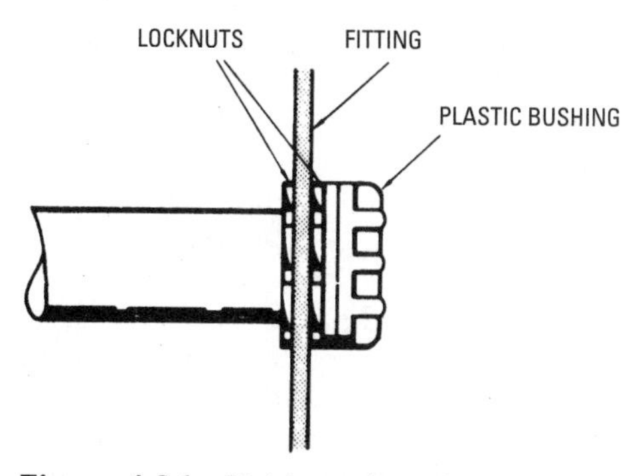

Figure 4-34 Rigid metal condult terminated at a junction box by locknuts and an insulated bushing.

344.42. Couplings and Connectors.

(A) Threadless. Threadless couplings and connectors used with conduit shall be made tight. Where buried in masonry or concrete, they shall be of the concretetight type. Where installed in wet locations, they shall be of the raintight type.

(B) Running Threads. Running threads shall not be used on conduit for connection at couplings.

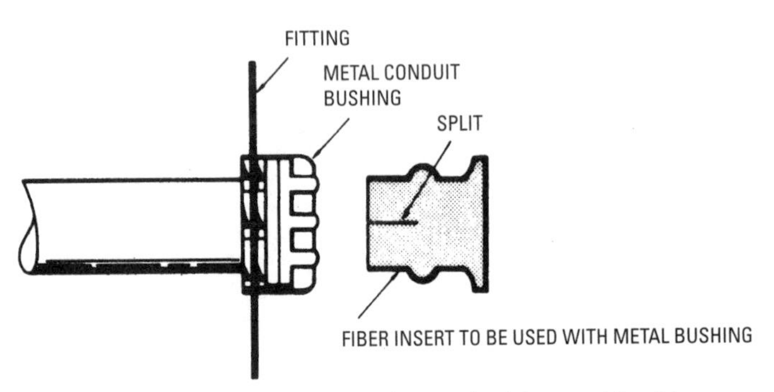

Figure 4-35 A fiber insert to be used with metal bushings.

For those who are not familiar with the term *running threads*, these were previously of a form that might be called a union, using water-pipe terminology. This was accomplished by threading the conduit back far enough so that a locknut could be screwed on in addition to a conduit coupling, which does not have tapered threading. Another piece of conduit was then screwed into the coupling, butting up against the first piece of conduit. This assembly was then locked by tightening the locknut that had previously been put on to the running thread (Figure 4-36).

Figure 4-37 illustrates a fitting that serves the same purpose on conduit that a union does on water pipe. It is often called an Erickson.

Pipe unions are not approved for conduit use. In fact, neither water pipe nor fittings are approved in place of rigid conduit. The interior coating of water pipe might be rough and damage the wire insulation. Attempts are often made to use water pipe and fittings for electrical raceways, even water-pipe sections welded together. When found, they have to be turned down and approved raceways and fittings used to replace them.

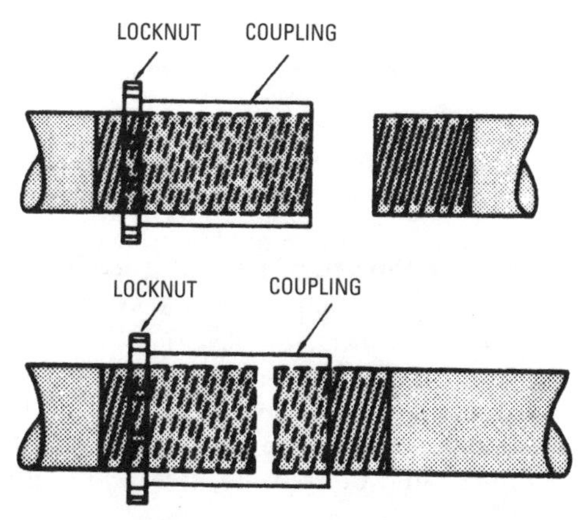

Figure 4-36 A conduit coupling using locknuts.

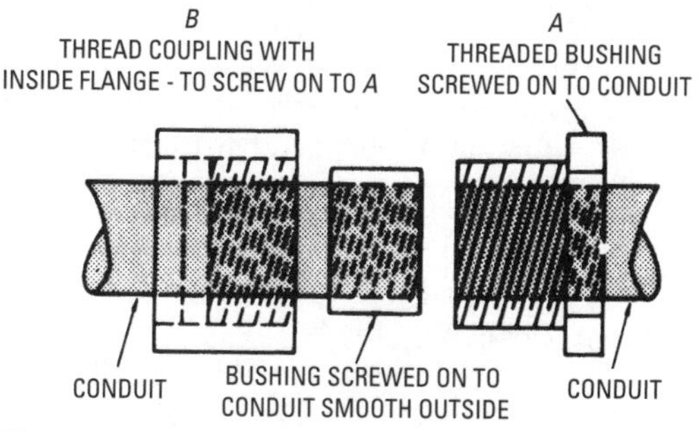

Figure 4-37 A type of condult coupling that serves as a union.

344.24. Bends—How Made. This was extensively covered in the section on EMT. Everything covered there is applicable to rigid metal conduit.

> *344.30. Securing and Supporting. Rigid metal conduit shall be installed as a complete system as provided in Article 300 and shall be securely fastened in place. Conduit shall be firmly fastened within 3 feet (914 mm) of each outlet box, junction box, cabinet, or fitting. Conduit shall be supported at least every 10 feet (3.05 m).*
>
> *Exception No. 1: If made up with threaded couplings, it shall be permissible to support straightruns of rigid metal conduit in*

accordance with Table 344.30(B)(2), provided such supports prevent transmission of stresses to termination where conduit is deflected between supports.

Notice that the *Code* specifies "threaded couplings." It definitely does not mention threadless couplings and connectors, discussed in *Section 344.42.*

Last, but by no means least, all connections into threaded hubs, locknuts, and bushings shall be made up tight. The rigid conduit is accepted as the equipment-grounding means, and if these are not made up tight, electrical continuity will not be good. Thus, in case of a ground fault, high impedance will result and thus impede the operation of the overcurrent devices. There may also be arcing at the connection to boxes, cabinets, and other enclosures, and there might be a resulting fire.

Boxes and Fittings

Boxes and fittings are covered in *Article 314* of the *NEC.* This chapter will discuss the usage of such boxes, how they shall be mounted, and the number of conductors permitted in each. Notice that cubic inch capacity is the criterion for the number and sizes of conductors permitted in boxes. Because these boxes are not always easily measured to obtain the cubic content, they are sometimes filled with water and the volume of the water measured instead.

Outlet or Switch Boxes: Types

There are quite a variety of boxes used in residential wiring. Perhaps the most common is the outlet or switch box, often referred to as a receptacle box. They are available in aluminum, galvanized steel, and plastic. Figure 4-38 illustrates some of these boxes.

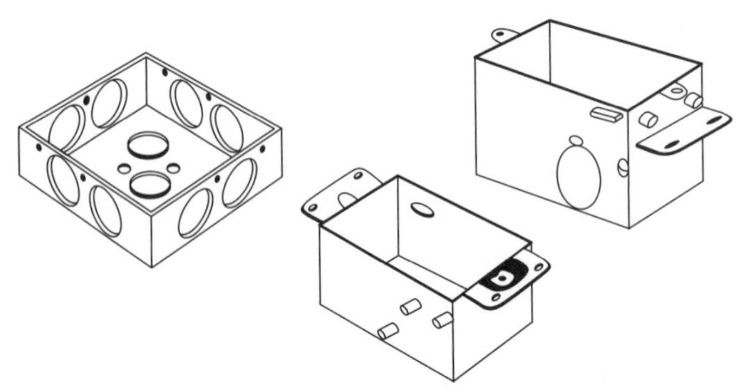

Figure 4-38 Various types of outlet or switch boxes.

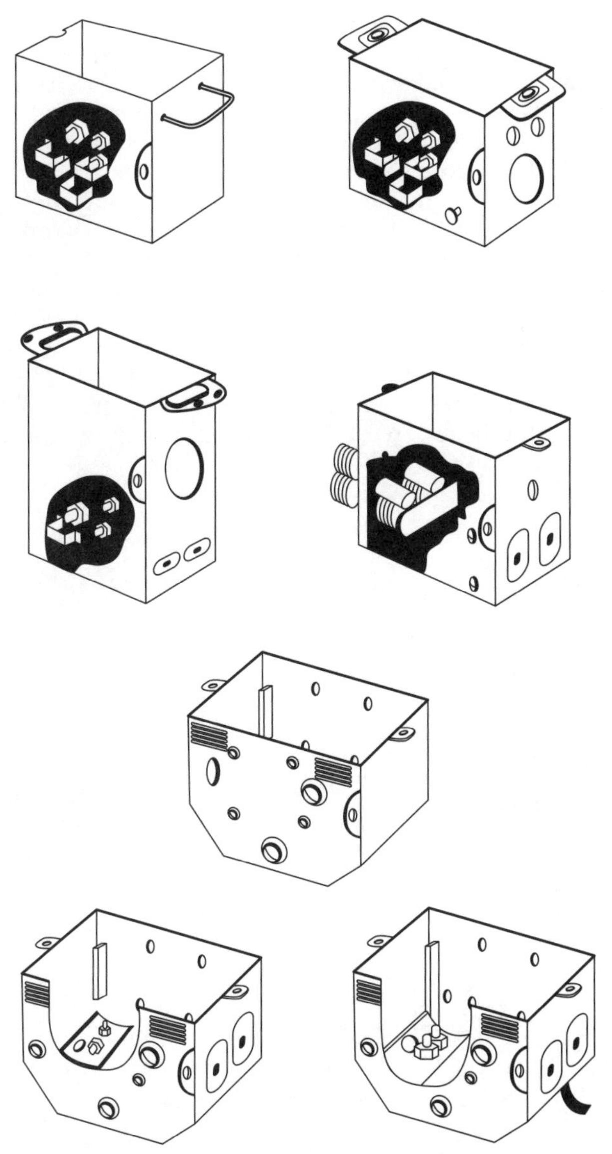

Figure 4-39 Various types of built-in cable clamps.

Some are for use with conduit, and this type is available with or without plaster ears, so it they may be mounted either in a plaster wall or in concrete and attached to rigid conduit or electrical metallic tubing. These same boxes may be used with armored cable or nonmetallic-sheathed cable by the use of approved clamps to secure these cables in the box. They are also available with clamps mounted in the box, one type approved for armored cable and another type approved for NM cable (Figure 4-39). Be certain that the boxes used are approved for the type of wiring being installed.

These boxes are available in single-gang types for mounting only one receptacle or switch, or for up to three switches and/or outlets of the Despard variety mounted in one single yoke. They are also available in multigang types for two or more receptacles or switches mounted horizontally in one row (Figure 4-40). Practically all of the single-gang boxes are of the sectional variety with which, by

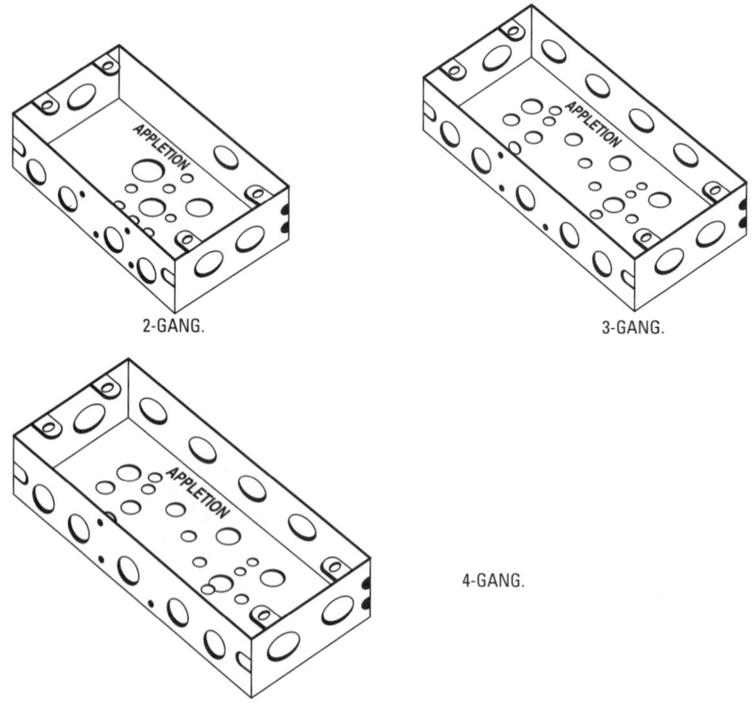

Figure 4-40 Ganged boxes.

removing one or both sides of two or more boxes, any size box can be assembled to allow the installation of multi-gang devices. Bakelite or plastic boxes must be purchased as ganged boxes in proper number, because they cannot be ganged in the field (Figure 4-41).

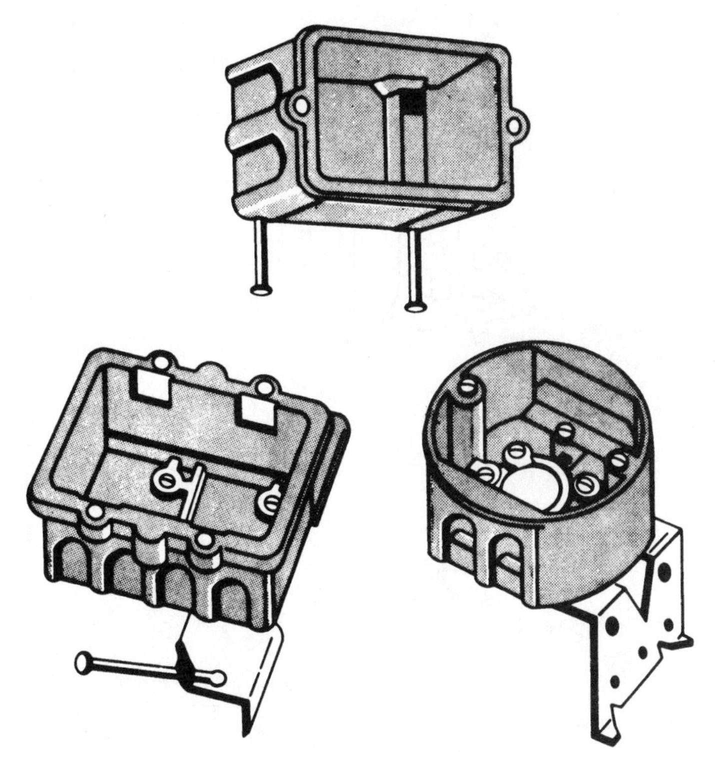

Figure 4-41 Plastic boxes.

Installation of Boxes

In using metal boxes, it is required that the cables be stapled within 12 inches of where they enter the box and that built-in clamps or approved connectors be used. Plastic boxes do not have built-in clamps for the cables, so it is therefore necessary to staple the cables within 8 inches of the box. In addition, NM cable is the only type permitted to use with plastic boxes.

There are available boxes with plaster ears, for mounting to wooden lath, but the majority are so called nail-on boxes, as shown in Figure 4-42.

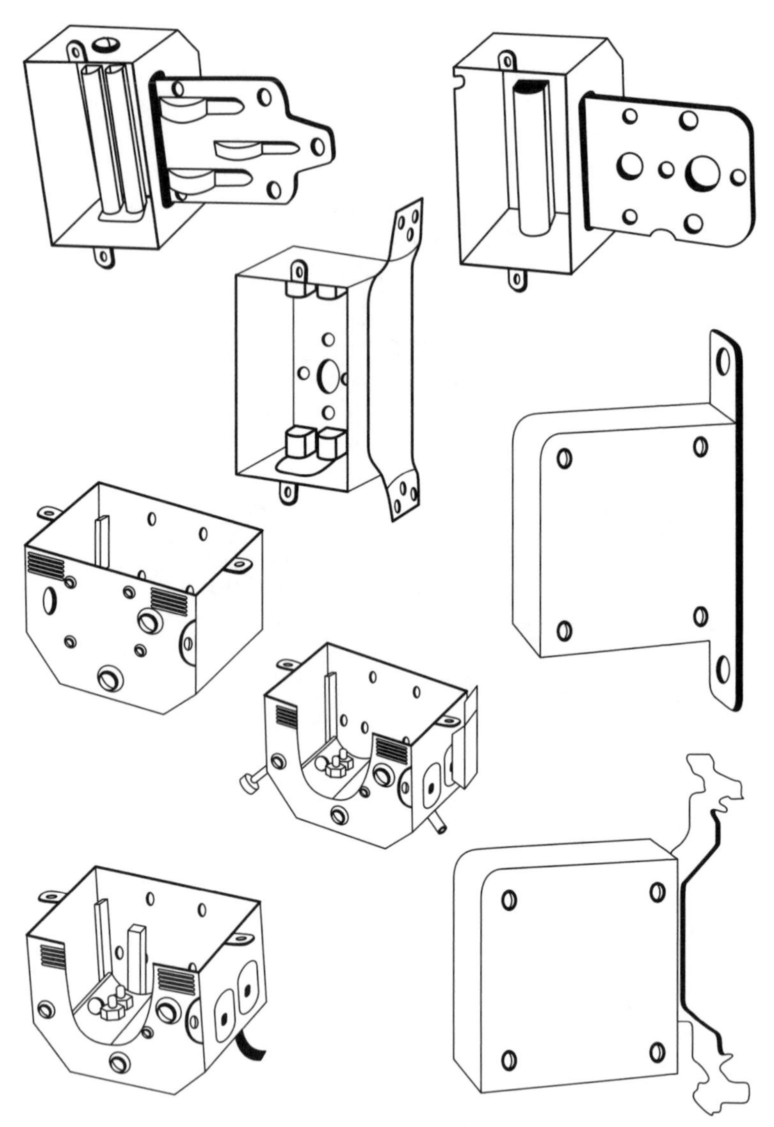

Figure 4-42 Various types of nail-on boxes.

Through-type boxes for mounting switches or receptacles on both sides of a wall are also manufactured. These are shown in Figure 4-43.

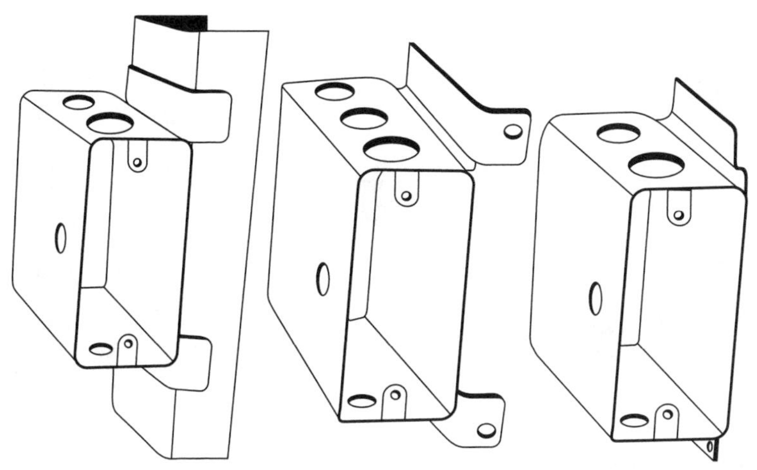

Figure 4-43 Through-type boxes for installing switches or receptacles on both sides of a wall.

Extensions are available that may be mounted on boxes to give a larger cubic capacity, for extending the box if it is mounted too far back, or for use if at some future date it is necessary to fur out the wall for paneling (Figure 4-44).

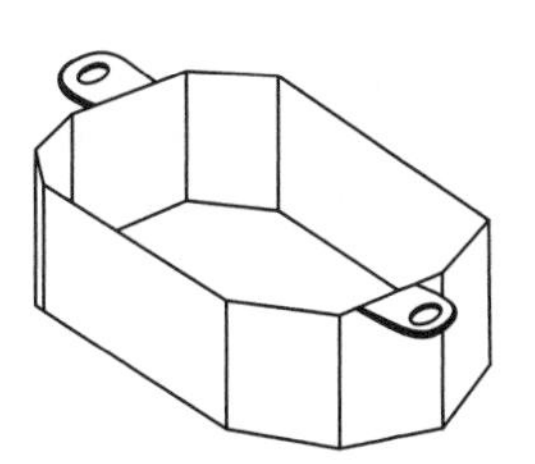

Figure 4-44 An extension for a switch box.

Where lath and plaster is used, the boxes usually become filled with plaster and hard to clean out, and they are sometimes hard to find as well. Threads must be cleaned after plastering. A cover for the protection of boxes in such installations is shown in Figure 4-45. This cover has a piece of gummed paper in the center with the notation, "Remove tape before plastering." The box is plastered over, but a small blue spot bleeds through the plaster and marks the position of the outlet. The plaster may then be removed, leaving a clean box that is well plastered in. Receptacle and switch boxes shall be mounted flush with all combustible materials, but may be set back

1/4 inch when installed in noncombustible materials. In all cases, however, the wall material must fit closely around the boxes.

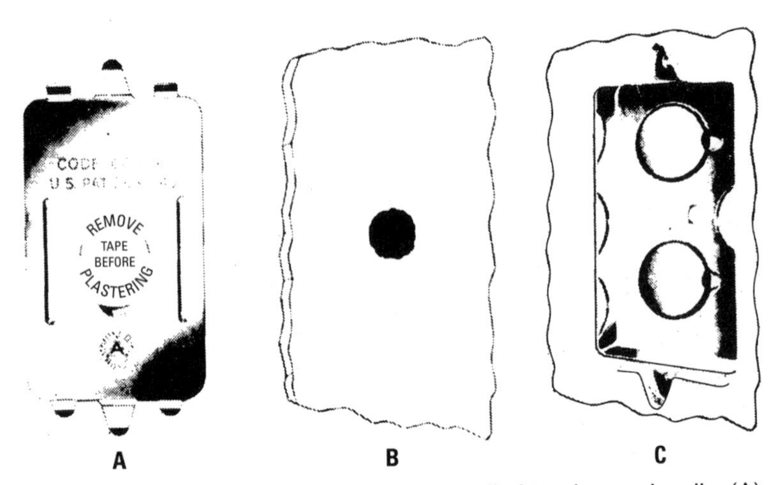

Figure 4-45 A cover for wall boxes installed in plastered walls: (A) cover with tape over color spot, (B) color spot bleeds through plaster to locate cover, and (C) plaster and cover removed, enabling access to box.

For mounting switches and receptacles, what is commonly known as a *4-inch square box* (Figure 4-46) is often used. Figure 4-46, part A shows a plain conduit type, but all of these boxes are available with clamps for NM cable or armored cable. Figure 4-46, parts B, C, and D illustrate 4-inch square boxes that may be attached to wooden studs without the use of nails. In the event a that 4-inch square box does not have enough capacity for the number and sizes of conductors that are being installed, extensions such as those shown in Figure 4-46, part E, may be added. One or more extensions may be used as required to obtain sufficient capacity. Note that 4-inch square boxes are not equipped for mounting receptacles or switches, so it is necessary to use what is known as a *plaster ring* (Figure 4-47). Although the term plaster ring is used, it does not mean that they are only to be used on plastered walls, since they are also usable for paneling or drywalls. They are available in different depths to accommodate the type of wall in which they are being installed.

Another type of box, called a *handy box*, is shown in Figure 4-48. These may be purchased with nail-on adaptors, but they are usually used for surface mounting. They do not come with clamps, so they

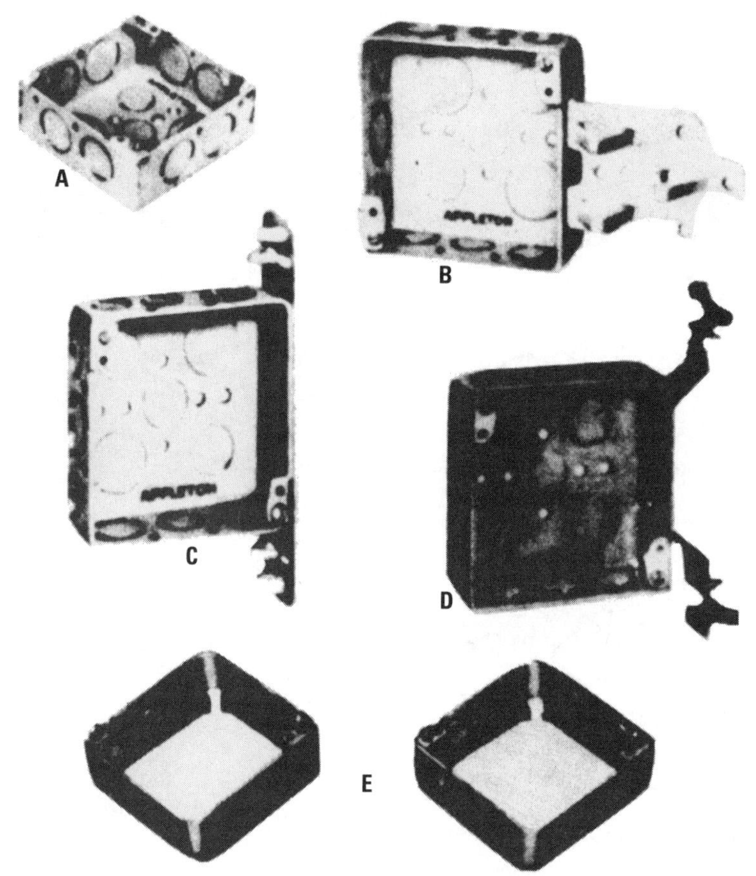

Figure 4-46 Various types of 4-inch square boxes.

are for use with conduit or EMT. If it is desired to use them with armored cable or NM cable (Romex), approved clamps must be used. Extensions are also available for handy boxes (Figure 4-49) to increase their capacity.

Figure 4-50 shows the method used to install a self-nailing 4-inch square box. The same method is used for self-nailing handy boxes and regular type receptacle boxes.

Figure 4-51 illustrates a 4-inch *octagon box* of the conduit or EMT type. This kind of box may also be used with cables, provided that approved clamps are added or that the box has approved built-in clamps (Figure 4-52).

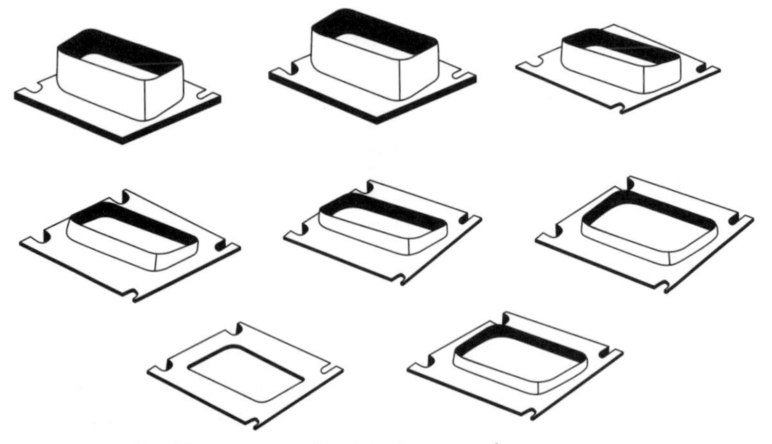

Figure 4-47 Plaster rings for 4-inch square boxes.

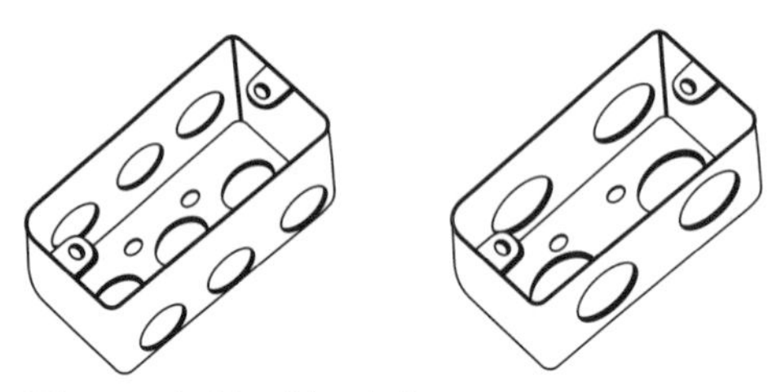

Figure 4-48 Handy boxes.

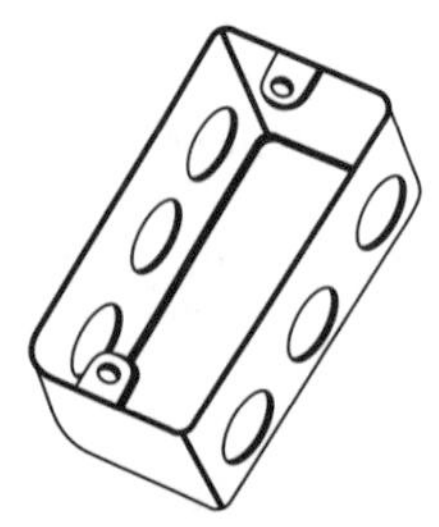

Figure 4-49 An extension for a handy box.

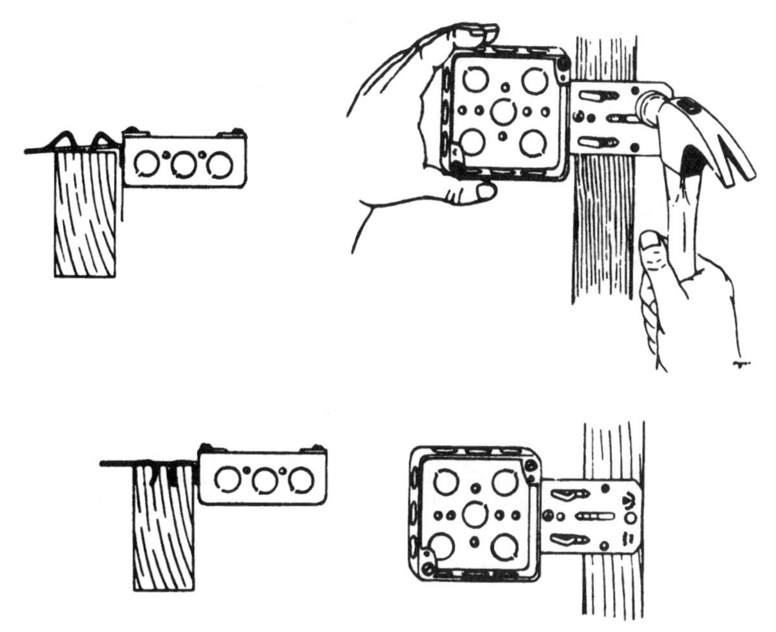

Figure 4-50 Method of installing self-nailing square boxes.

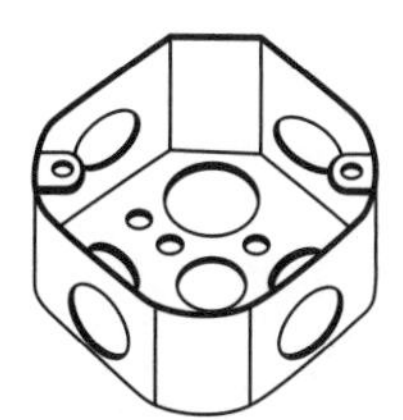

Figure 4-51 A plain 4-inch octagon box.

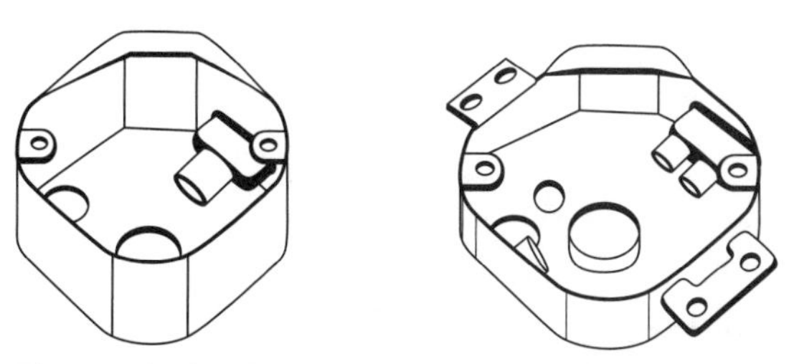

Figure 4-52 Four-inch octagon boxes with built-in cable clamps.

Figure 4-53 shows 4-inch octagon boxes of the nail-on type. Extension rings (Figure 4-54) are available for 4-inch octagon boxes to gain sufficient capacity for the number and sizes of conductors being used.

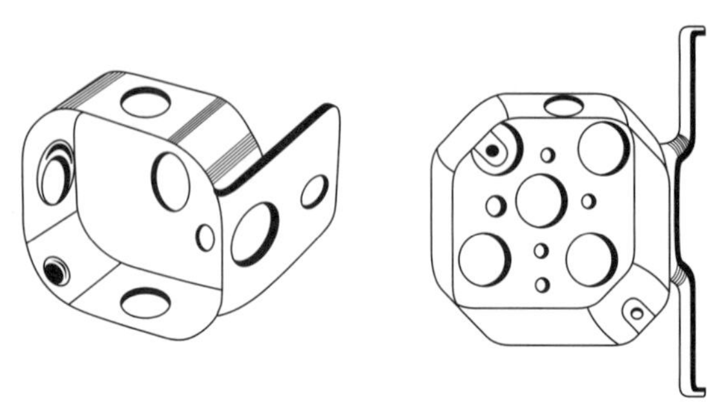

Figure 4-53 Nail-on type octagon boxes.

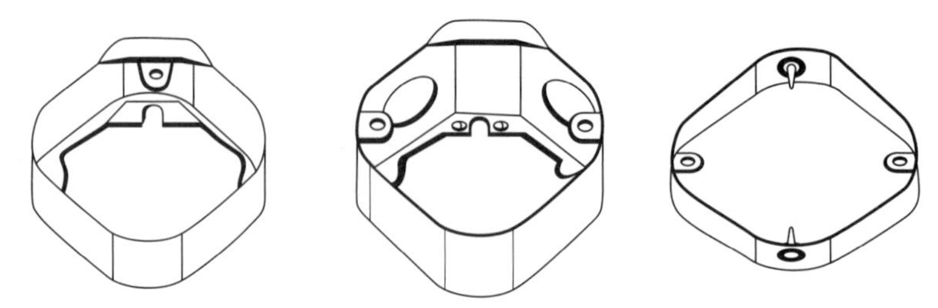

Figure 4-54 Extension rings for octagon boxes.

If the walls or ceilings are plastered or if drywall is used, 4-inch octagon boxes should have plaster rings of a type similar to those shown in Figure 4-47. This type of box may be mounted with mounting brackets similar to those shown for a common box or a 4-inch square box. It is also possible to install a 2-by-4 between the ceiling joists and mount the octagon box on this. There are, however, many and varied types of patented hangers available, two of which are shown in Figure 4-55.

Figure 4-56 shows a plaster ring installed on a 4-inch square box. Remember, these rings are available in various depths, so purchase the one required.

Number of Conductors

Section 314.16 is a very vital portion of *Article 314* concerning the number of conductors allowed in a box. This is of such vital

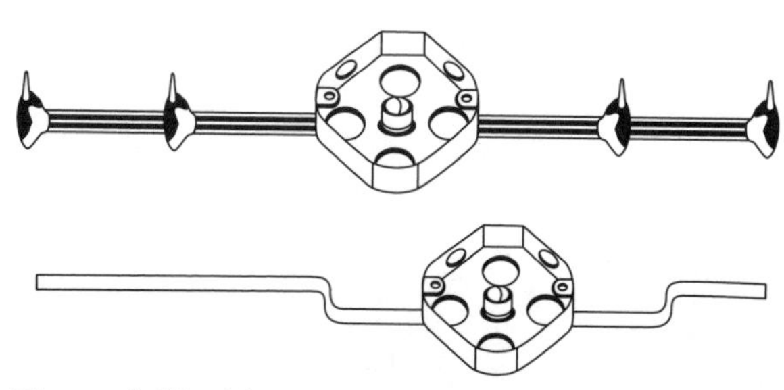

Figure 4-55 Mounting brackets for octagon boxes.

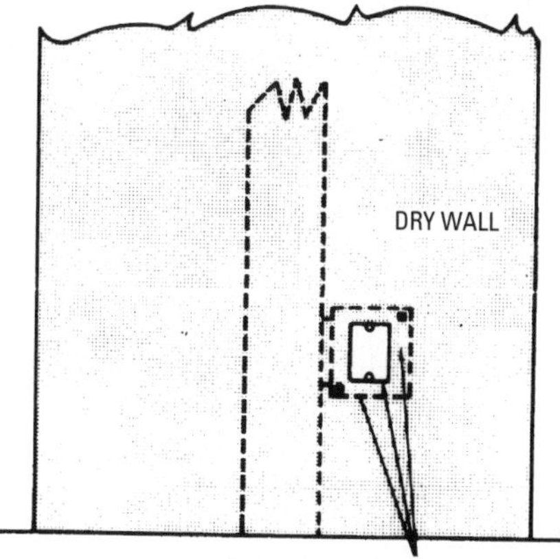

Figure 4-56 A plaster ring installed on a 4-inch square box.

importance that several portions of *Section 314.16* and *Tables 314.16(A)* and *314.16(B)* are reproduced here along with any necessary comments.

> *314.16. Number of Conductors in Outlet, Device, and Junction Boxes, and Conduit Bodies. Boxes shall be of sufficient size to provide free space for all conductors. In no case shall the volume of the box , as calculated in 314.16(A), be less than the fill calculation as calculated in 314.16(B). The minimum volume for conduit bodies shall be as calculated in 314.16(C). The provisions of this section shall not apply to terminal housings supplied with motors.*

Boxes and conduit bodies containing conductors, size No. 4 or larger, shall also comply with the provisions of Section 314.28.

(A)(1) Standard Boxes. The volumes of standard boxes that are not marked with their volume shall be as given in Table 314.16(A).

The statement "Boxes shall be of sufficient size to provide free space for all conductors" is a basic and broad statement that, in itself, is quite sufficient; however, a complete analysis of this will be made here, as it is in the *Code*. The main point is that in the installation of conductors in boxes, the conductors should never be forced into the box, as this is a potential source of trouble. The *Code* spells out what is good practice as well as the minimum requirements. The installer should, however, always bear in mind the intent and, if necessary to do a good job, go even further than the minimum requirements.

The limitations imposed by *Sections 314.16(A)* and *(B)* are not intended to apply to terminal housings supplied with motors, nor to those types of boxes or fittings without knockouts that have hubs or recessed parts for terminal bushings and locknuts.

The maximum numbers of conductors listed in *Table 314.16(A)*, reproduced here as *Table 4-6*, apply without change where no fittings or devices such as fixture studs, cable clamps, hickeys, switches, or receptacles are contained in the box and where no grounding conductors are part of the wiring within the box. Where one or more such devices are contained in the box, the number of conductors shown in the table must be reduced, as will be described in subsequent paragraphs.

The volume of a wiring enclosure (box) shall be the total volume of the assembled sections, and, where used, the space provided by plaster rings, domed covers, extension rings, and the like, that are marked with their volume in cubic inches or are made from boxes whose dimensions are listed in *Table 4-6*.

For combinations of conductor sizes shown in *Table 4-6*, the volume per conductor listed in *Table 314.16(B)*, reproduced here as *Table 4-7*, applies. The maximum number and size of conductors listed in *Table 4-7* shall not be exceeded.

314.16(A)(2) Other Boxes. Boxes 100 cubic inches or less other than those described in Table 314.16(A), and nonmetallic boxes shall be durably and legibly marked by the manufacturer with their volume. Boxes described in Table 314.16(A) that have a larger cubic inch capacity than is designated in

> *the table shall be permitted to have their cubic inch capacity marked as required by this section.*
>
> *314.16(C) Conduit Bodies. Conduit bodies enclosing No. 6 conductors or smaller other than short-radius conduit bodies as described in 314.15, shall have a cross-sectional area not less than twice the cross-sectional area of the largest conduit or tubing to which it is attached. The maximum number of conductors permitted shall be the maximum number permitted by Table 1, Chapter 9 of the NEC, for the conduit or tubing to which it is attached.*

Table 4-6 covers the maximum number of conductors that will be permitted in outlet and junction boxes. As previously mentioned, there is been no allowance in this table for fittings or devices such as studs, cable clamps, hickeys, switches, or receptacles that are contained in the box. These must be taken into consideration and deductions made for them. A deduction of *two* conductors shall be made for each of the following:

1. Fixture studs
2. Cable clamps
3. Hickeys

A deduction of *one* conductor shall be made for each of the following:

1. Grounding conductors
2. Conductors that run all the way through the box
3. Conductors originating outside the box and terminating inside the box

Conductors of which no part leaves the box will not be counted, such as wires to fixtures that are spliced onto the other wires in the box.

Boxes are often ganged together with more than one device per strap mounted in these ganged boxes. In these cases, the same limitations will apply as if they were individual boxes. Please refer to Figures 4-57 through 4-60 for illustrations concerning this section.

Installation of Conductors

The question concerning plaster rings and the effect that they have on the number of conductors permitted in the box often arises. Although the plaster ring is meant for another purpose, almost every inspector will agree that it provides extra space for conductors

Table 4-6 Metal Boxes, *Table 314.16(A)* in the *NEC*

Box Dimensions, Inches Trade Size or Type	*Min. Cu. In. Cap.*	*Maximum Number of Conductors*						
		No. 18	*No. 16*	*No. 14*	*No. 12*	*No. 10*	*No. 8*	*No. 6*
4 × 1 1/4 Round or Octagonal	12.5	8	7	6	5	5	5	2
4 × 1 1/2 Round or Octagonal	15.5	10	8	7	6	6	5	3
4 × 2 1/8 Round or Octagonal	21.5	14	12	10	9	8	7	4
4 × 1 1/4 Square	18.0	12	10	9	8	7	6	3
4 × 1 1/2 Square	21.0	14	12	10	9	8	7	4
4 × 2 1/8 Square	30.3	20	17	15	13	12	10	6
4 11/16 × 1 1/4 Square	25.5	17	14	12	11	10	8	5
4 11/16 × 1 1/2 Square	29.5	19	16	14	13	11	9	5
4 11/16 × 2 1/8 Square	42.0	28	24	21	18	16	14	8
3 × 2 × 1 1/2 Device	7.5	5	4	3	3	3	2	1
3 × 2 × 2 Device	10.0	6	5	5	4	4	3	2
3 × 2 × 2 1/4 Device	10.5	7	6	5	4	4	3	2
3 × 2 × 2 1/2 Device	12.5	8	7	6	5	5	4	2

Table 4-6 (continued) Metal Boxes, *Table 314.16(A)* in the *NEC*

Box Dimensions, Inches Trade Size or Type	Min. Cu. In. Cap.	Maximum Number of Conductors						
		No. 18	No. 16	No. 14	No. 12	No. 10	No. 8	No. 6
3 × 2 × 2 3/4 Device	14.0	9	8	7	6	5	4	2
3 × 2 × 3 1/2 Device	18.0	12	10	9	8	7	6	3
4 × 2 1/8 × 1 1/2 Device	10.3	6	5	5	4	4	3	2
4 × 2 1/8 × 1 7/8 Device	13.0	8	7	6	5	5	4	2
4 × 2 1/8 × 2 1/8 Device	14.5	9	8	7	6	5	4	2
3 3/4 × 2 × 2 1/2 Masonry Box/Gang	14.0	9	8	7	6	5	4	2
3 3/4 × 2 × 3 1/2 Masonry Box/Gang	21.0	14	12	10	9	8	7	4
FS-Minimum Internal Depth 1 3/4 Single Cover/Gang	13.5	9	7	6	6	5	4	2
FD-Minimum Internal Depth 2 3/8 Single Cover/Gang	18.0	12	10	9	8	7	6	3
FS-Minimum Internal Depth 1 3/4 Multiple Cover/Gang	18.0	12	10	9	8	7	6	3
FD-Minimum Internal Depth 2 3/8 Multiple Cover/Gang	24.0	16	13	12	10	9	8	4

Table 4-7 Volume Required per Conductor, *Table 314.16(B)* in the *NEC*

Size of Conductor (No.)	*Free Space Within Box for Each Conductor (cu. in.)*
18	1.5
16	1.75
14	2.0
12	2.25
10	2.5
8	3.0
6	5.0

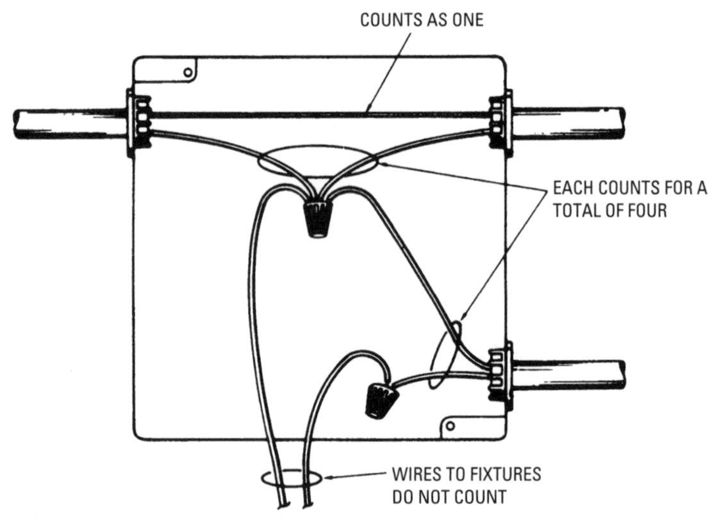

Figure 4-57 Which wires to count in a junction box.

and that the conductor allowance for the box should be raised accordingly. However, pushing the allowance so tightly isn't a good idea, and it is generally considered best not to include ring capacity in box fill calculations. The entire intent is not to crowd the conductors and thereby cause failure of the insulation.

Care must be exercised in protection of the conductors from abrasion where they enter the boxes or fittings. With conduit, this is

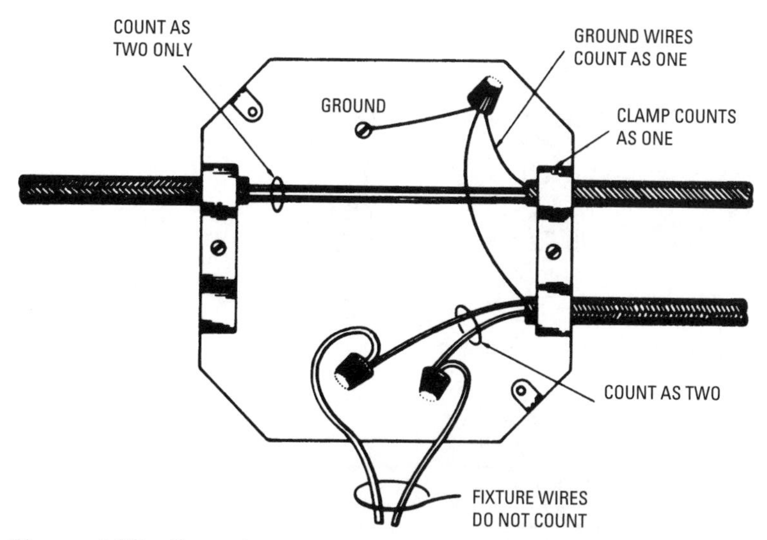

Figure 4-58 Grounding wire counts as one conductor.

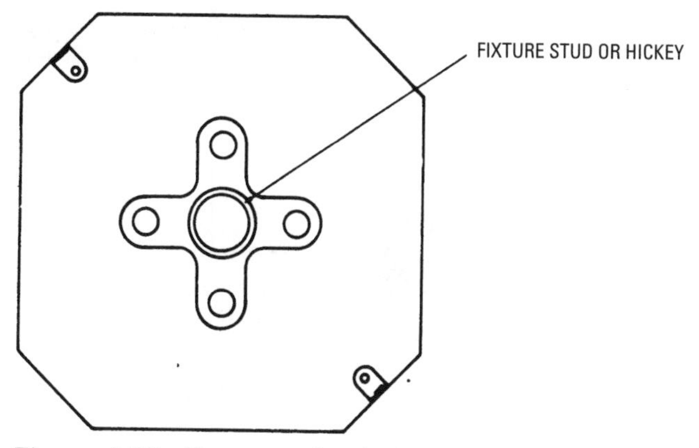

Figure 4-59 Fixture stud or hickey counts as two conductors.

accomplished by bushings or other approved devices. With NM cable, as may be seen in Figure 4-61, the outer covering of the cable should protrude from the clamp to provide this protection. With armored cable, fiber bushings are to be inserted between the conductors and the armor to prevent any abrasion.

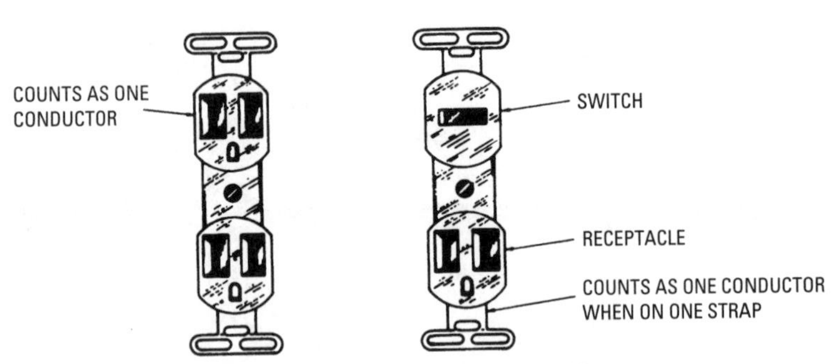

Figure 4-60 How to count devices in figuring fill.

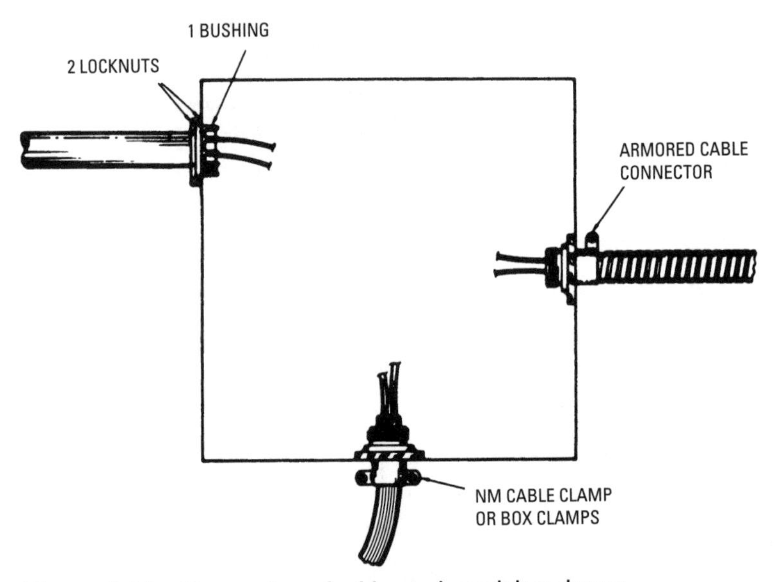

Figure 4-61 Connection of cables and condult to boxes.

The following rules shall be complied with:

- Openings through which conductors enter must be adequately closed. When cables enter a box, cable clamps shall be used or the boxes provided with built-in cable clamps; with conduit, the locknuts and bushings will adequately close the openings.
- When metal boxes or fittings are used with open wiring, proper bushings must be used. In dry places, flexible tubing may be

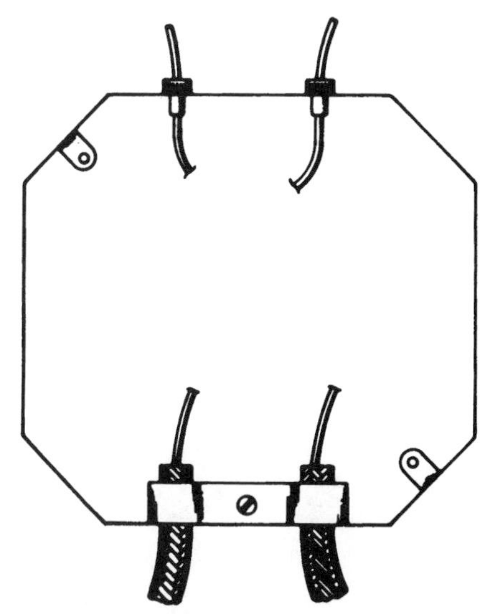

Figure 4-62 Open wiring into boxes.

used and extended from the last conductor support into the box and secured (Figure 4-62).

- Where nonmetallic boxes are used with either concealed knob-and-tube work or open wiring, the conductors must pass through individual holes in the box. If flexible tubing is used over the conductor, it shall extend from the last conductor support into the hole in the box.
- Where nonmetallic cable is used, it must extend through the opening in the box. It is not required that individual conductors or cables be clamped if the individual conductors or cables are supported within 8 inches of the box. When nonmetallic conduit is used with nonmetallic boxes, the conduit shall be connected to the box by approved means.

Chapter 5

Wiring for Electrical House Heating

Electrical heating is used in a great number of homes. The secret of good, economical house heating by means of electricity lies in following certain tried-and-true methods of installing insulation and vapor barriers in an accepted manner. The figuring of a house heating system is another subject; if you are not fully qualified on this subject, consult your local utility company; they should have qualified engineers for advice and calculations. They are interested in low-cost heating and will lay out the job on the most economical basis. In the past, some utility companies have supplied the materials at slightly below cost in order to increase their consumer listing.

We begin our coverage in this chapter with the underlying environmental issues. We will then cover the rules of installation from an electrical standpoint.

Environmental Issues

In designing and installing heating systems, nonelectrical factors must be considered. In these applications, the end goal is not just a system that functions properly from an electrical standpoint but one that provides efficient, affordable, and comfortable heating.

The two primary issues are the house's insulation and the control of humidity. Insulation plays a huge role in the affordability of heating systems, and humidity plays a large role in comfort.

The insulation that you will use is vital. Many electrical contractors who install heating also install insulation and, in doing so, save themselves many problems that might arise due to an improperly insulated home. This would cause high electric bills that are entirely unnecessary. The insulating factor and the thickness of the insulation play an important part in the cost of heating. In a well-insulated home, the humidity will rise, which adds to comfort in the winter. With the higher humidity, a vapor barrier must be installed between the insulation and the drywall or lathing and plaster, because moisture will cut the efficiency of the insulation.

It might be well to consider a humidistat control system to keep the humidity at the proper level. This would be especially true in damp climates. The major heat losses are through windows, doors, and ceilings, so attention must be paid to these items. The construction of the house has no place in this book but is mentioned to help you fulfill your job well.

There are many types of electrical heating systems, such as:

1. Ceiling heat cables
2. Central heating systems
3. Baseboard heating, which may consist of conventional resistive heating units or baseboard water-type heating units
4. Heating panels
5. Unit heaters
6. Heat pumps

Electrical Protection Issues

The primary risk pertaining to electrical heating is overheating, possibly leading to fire. This is avoided through the use of proper materials and the proper use of overcurrent devices.

The sizing of overcurrent devices and branch-circuit conductors for electrical (fixed) space heating should be calculated on the basis of 125 percent of the total load of the heaters and motors (if equipped with motors) and 125 percent of the total load of the heaters, if not equipped with motors.

For example, if a heating circuit draws 14 amperes full load, the circuit conductors and the overcurrent devices will have to be sized to a minimum of 125 percent of 14 amperes, or $17\frac{1}{2}$ amperes. So a 20-ampere breaker or a fuse will be required as well as No. 12 copper conductors or No. 10 aluminum or copper-clad aluminum conductors.

Usually each room has its own thermostatic control. Some thermostats have an "off" position, and some do not. Where there is an "off" position, the thermostat switch must break all hot conductors (not the neutral) simultaneously (Figure 5-1).

A very common method of heating homes is by means of ceiling heat cable. This is a very effective means of heating (and quite trouble-free) if installed properly. One might ask, "What happens if a cable opens after the installation has been in operation for some time?" There is "open cable" locating equipment available, and the ceiling may be opened at that point and repairs made, but make sure that the proper equipment is used in making the repairs. Figure 5-2 illustrates ceiling heat cable construction. Note the 7 feet of nonheating leads. These *shall not be shortened*—the ends have the markings required by the *NEC* and UL. The heat cable may be identified by the markings at the ends of the nonheating leads and

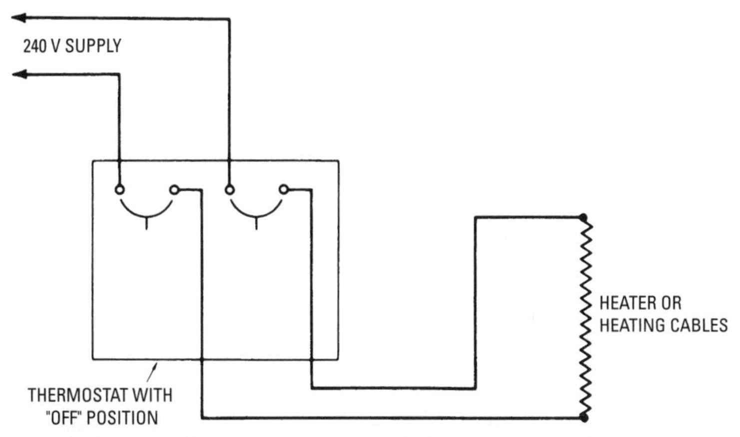

Figure 5-1 A thermostat with an "off" position shall disconnect all hot supply lines.

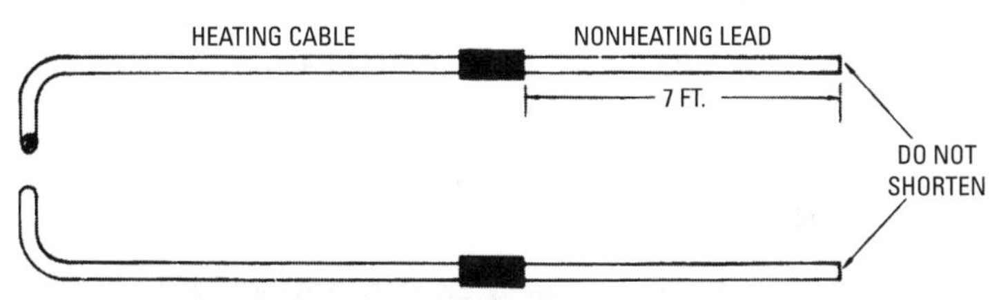

Figure 5-2 Nonheating leads shall be a minimum of 7 feet.

by the color as shown in *Table 5-1* (there are other colors for other voltages, but these are the voltages that you find in homes). Be sure the cables used carry the UL listing mark.

Installation of Electrical Heating

Before we go through all of the electrical installation issues related to electrical heating, it is important to state that the installation of these heating systems must frequently be coordinated with the work of other trades. This is not necessarily true of baseboard and unit heaters, but it is definitely an issue for heating cables that are installed in ceilings or floors.

Special coordination is required for any heating elements that will be embedded in plaster, concrete, or similar surfaces. The following

Table 5-1 Heating Cable Color Code

120-volt nominal voltage	Yellow
208-volt nominal voltage	Blue
240-volt nominal voltage	Red

installation rules must be followed, but all your work will have to be performed in cooperation with other trades:

> *424.38. Area Restrictions.*
> *(A) Shall Not Extend Beyond the Room or Area. Heating cables shall not extend beyond the room or area in which they originate.* [See Figure 5-3.]
>
> *(B) Uses Prohibited. Cables shall not be installed in closets, over walls or partitions that extend to the ceiling, or over cabinets whose clearance from the ceiling is less than the minimum horizontal dimension of the cabinet to the nearest cabinet edge that is open to the room or area.*
>
> *Exception: Isolated runs of cable may pass over partitions where they are embedded.*

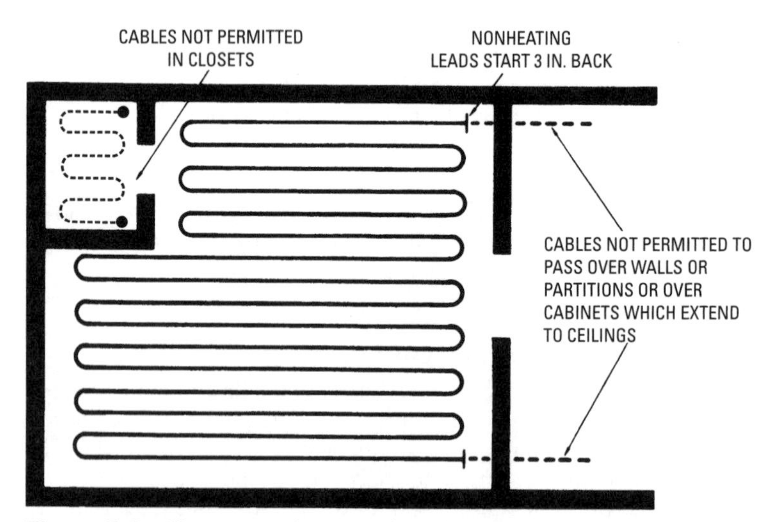

Figure 5-3 Illustration showing where heating cables may and may not be installed in ceilings.

(C) In Closet Ceilings as Low Temperature Heat Sources to Control Relative Humidity.

There are climates where humidity control is required in closets, but the prohibition of cables in closets is not intended to forbid the use of low-temperature humidity controls in closets (Figure 5-3).

424.39. Clearance from Other Objects and Openings. Heating elements of cables shall be separated at least 8 inches (203 mm) from the edge of outlet boxes and junction boxes that are to be used for mounting surface lighting fixtures. A clearance of not less than 2 inches (50.8 mm) shall be provided from recessed fixtures and their trims, ventilation openings, and other such openings in room surfaces. Sufficient area shall be provided to assure that no heating cable will be covered by any surface-mounted units.

The temperature limits and overheating of the cables are involved in this instance. Therefore, the requirements of this section should be followed very carefully, and, in cases of doubt, a little extra clearance should be given (Figure 5-4).

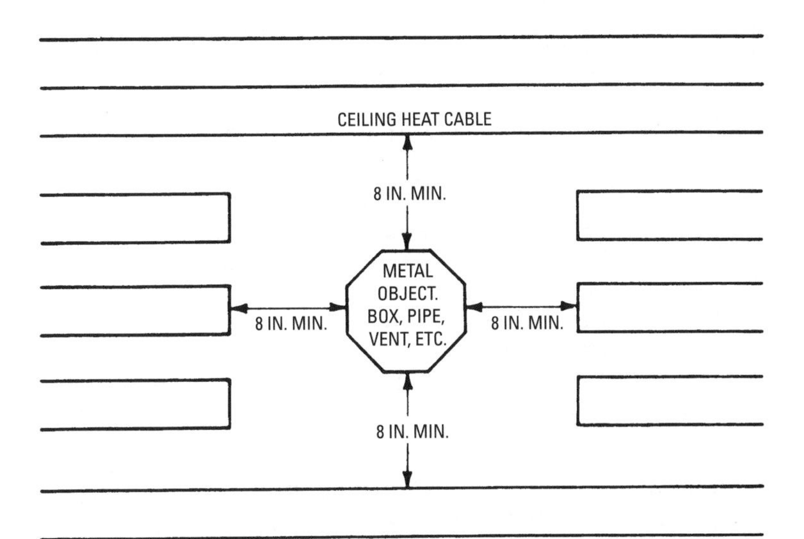

Figure 5-4 Clear metal heating cable from metal objects by a minimum of 8 inches.

424.40. *Splices.* Splicing of cables is prohibited except where necessary due to breaks. Even then the length of the cable should not be altered, because doing so will change the characteristics of the cable and the heat. It will occasionally be necessary to splice a break, but only approved methods will be used.

424.41. *Installation of Heating Cables on Dry Board, in Plaster, and on Concrete Ceilings.*

> *(A) Shall Not Be Installed in Walls. Heating cable shall not be installed in walls. It is not designed for this purpose and is strictly forbidden, with the exception that isolated runs of cable may be run down a vertical surface to reach a drop ceiling.*
>
> *(B) Adjacent Runs. Adjacent runs of heating cable shall be spaced not closer than $1^1/_2$ inches on centers and have wattage not to exceed $2^3/_4$ watts per foot.* [See Figure 5-5.]

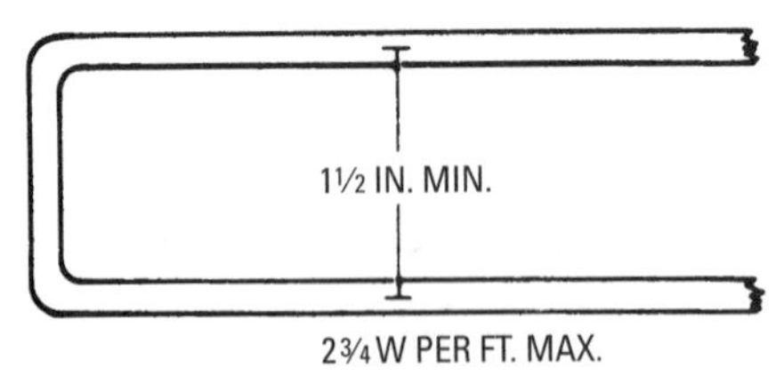

Figure 5-5 Heating cable installed in ceilings shall be spaced at least $1^1/_2$ inches apart.

> *(C) Surfaces to be Applied.* Heating cables shall be applied only to gypsum board, plaster lath, or similar fire-resistant materials. If cable is applied to the surface of metal lath or any other conducting material, there shall be a coat of plaster, commonly known as a brown or scratch coat, applied before the cable is installed. This coating of plaster shall entirely cover the metal lath or conducting surface. See *Section 424.41(F)* below.
>
> *(D) Splices. All heating cables, the splice between the heating cable and nonheating leads, and 3-inch (76 mm) minimum of the nonheating lead at the splice shall be embedded in plaster or dry board in the same manner as the heating cable.* [See Figure 5-3.]

(E) Ceiling Surfaces. On plastered ceilings, the entire surface shall have a finish coat of thermally noninsulating sand plaster or other approved coating that shall have a nominal thickness of ½ inch. Insulation (thermal) plaster shall not be used (Figure 5-6).

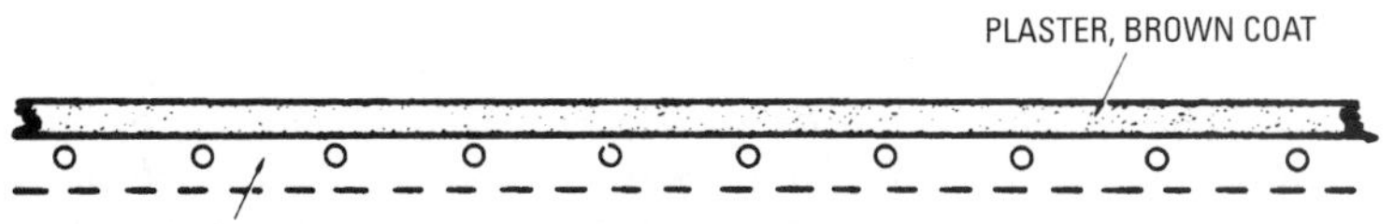

Figure 5-6 Heating cable installed in plaster ceiling.

(F) Secured. The cable shall be fastened at intervals not to exceed 16 inches by means of taping, stapling, or plaster. Staples or metal fasteners that straddle the cable shall not be used with metal lath or other conducting surfaces. The fastening devices shall be of an approved type.

(G) Dry Board Installations. When dry board ceilings are used, the cable shall be installed and the entire ceiling below the cable shall be covered with gypsum board not exceeding ½ inch in thickness, but voids between the two layers and around the cable shall be filled with a conducting plaster or other approved thermal conducting material so that the heat will readily transfer (Figure 5-7).

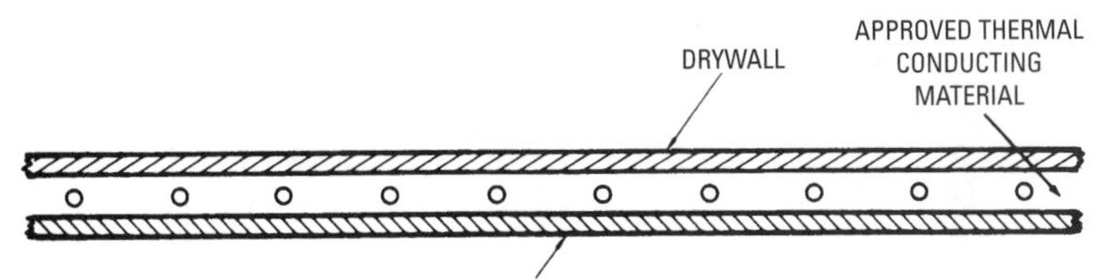

Figure 5-7 Heating cable installed in drywall ceiling.

(H) Free from Contact with Conductive Surfaces. Heating cables shall not come in contact with metal or other conducting materials.

(J) Joists. In dry board applications, cable shall be installed parallel to the joist, leaving a clear space centered under the

joist of $2^1/_2$ inches (64 mm) between centers of adjacent runs of cables. Crossing of joist by cable shall be kept to a minimum. Surface layer of gypsum board shall be mounted so that the nails or other fasteners do not pierce the heating cable. Where practicable, cables shall cross joists only at the ends of a room.

424.42. Finished Ceilings. The question often arises as to whether wallpaper or paint can be used over a ceiling that has heating cable. These materials have been used as finishes over heating cable since cables were first used in the late 1940s. This section gives formal recognition to painting or papering ceilings.

424.43. Installation of Nonheating Leads of Cables.

(A) Free Nonheating Leads. Only approved wiring methods shall be used for installing the nonheating leads of cables or panels from junction boxes to the underside of the ceiling. In these installations, single leads in raceways (conductors) or single or multiconductor Type UF, Type NMC, Type MI, or other approved conductors shall be used. Please note the absence of Type NM.

(B) Leads in Junction Box. Where nonheating leads terminate in a junction box, there shall be not less than 6 inches of nonheating leads free within the junction box. Also, the markings of the leads shall be visible in the junction box. This is highly important so that the heating cable can be identified (Figure 5-8).

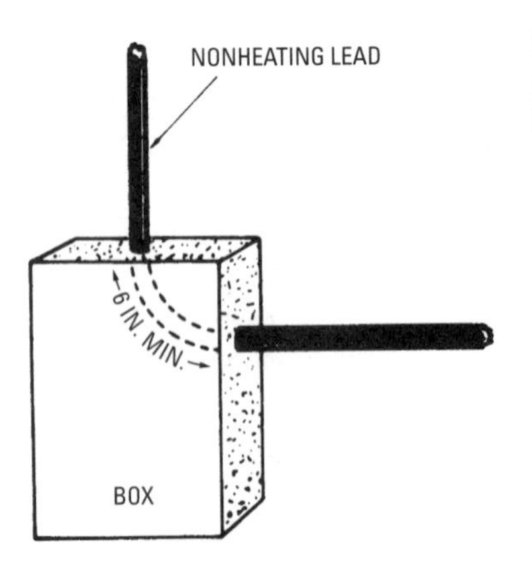

Figure 5-8 Installing nonheating leads in a junction box.

(C) Excess Leads. Excess leads of heating cables shall not be cut but shall be secured to the underside of the ceiling and embedded in plaster or other approved material, leaving only a length sufficient to reach the junction box with not less than 6 inches (152 mm) of free lead within the box. [See Figure 5-8.]

424.44. Installation of Panels or Cables in Concrete or Poured Masonry Floors. This section is for fixed indoor space heating and is not to be confused with ice and snow melting.

(A) Watts per Linear Foot. Panels or heating units shall not exceed 16½ watts per linear foot of cable.

(B) Spacing Between Adjacent Runs. The spacing between adjacent runs of cable shall not be less than 1 inch (25.4 mm) on centers.

(C) Secured in Place. Cables have to be secured in place while concrete or other finish material is being applied. Approved means, such as nonmetallic spreaders of frames, shall be used. Concrete floors often have expansion joints in them. Cables, units, and panels shall be so installed that they do not bridge an expansion joint unless they are protected so as to prevent damage to the cables, units, or panels due to expansion or contraction of the floor.

(D) Spacings Between Heating Cable and Metal Embedded in the Floor. Spacings shall be maintained between the heating cable and metal embedded in the floor, unless the cable is grounded metal-clad cable. This includes MI heating cable, which is extensively used.

(E) Leads Protected. The nonheating leads shall be sleeved with rigid metal conduit, intermediate metal conduit, rigid nonmetallic conduit, electrical metallic tubing, or other approved means for protection where the leads leave the floor (Figure 5-9).

(F) Bushings or Approved Fittings. To prevent damage to the cable, the sleeves mentioned in *(E)* shall have bushings or other approved means used where the leads enter or emerge from the floor slab.

Wiring above heated ceilings shall be located not less than 2 inches above the ceiling. If it is in thermal insulation at this height, it shall

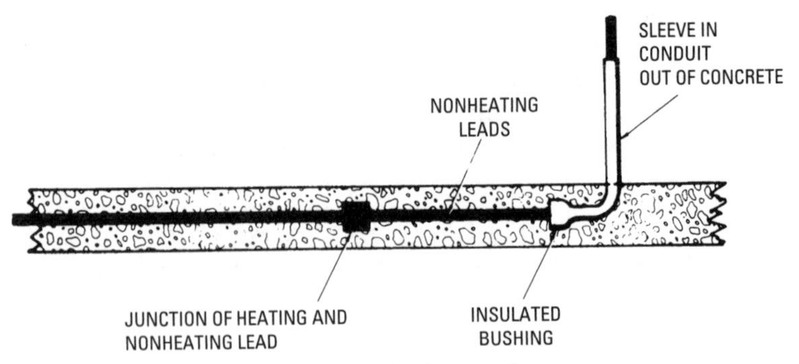

Figure 5-9 Sleeving nonheating leads out of concrete.

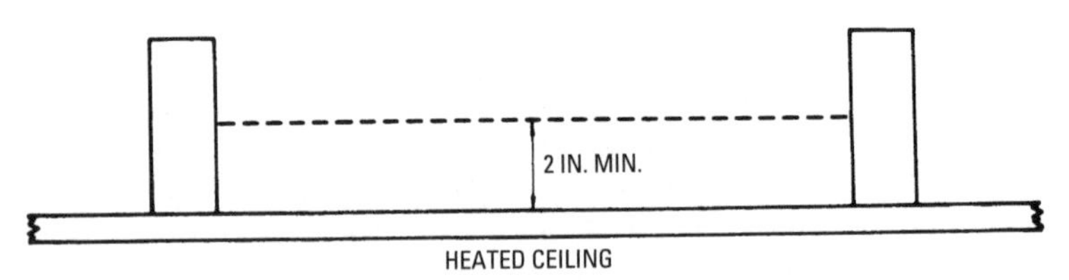

Figure 5-10 Conductors shall not be less than 2 inches above heated ceiling.

be considered as operating at 50°C (122°F), and no special protection will be required (Figure 5-10).

When ready to install heat cable, draw the room out to scale on paper, get the footage of the cable to be installed, and lay it out on the drawing so that you can evenly space the cable to use all of it. Then lay out the drawing on the ceiling with a chalk line.

Chapter 6

Mobile Homes

A mobile home is actually a residence, but because of its mobility the structure has different requirements from a regular residential dwelling. *Section 550.2* defines a mobile home as follows:

> *A factory-assembled structure or structures equipped with the necessary service connections and made so as to be readily movable as a unit or units on its own running gear and designed to be used as a dwelling units(s) without a permanent foundation.*

The phrase "without a permanent foundation" indicates that the support system is constructed with the intent that the mobile home placed thereon will be moved from time to time at the convenience of the owner.

Wiring for mobile homes generally follows the same requirements as those for other homes, though there are some significant differences. The material we will cover in this chapter focuses primarily on these differences.

We are not covering wiring *inside* mobile homes. There are differences between field-installed wiring (such as in a traditional house) and factory-installed wiring (such as in a mobile home). But the special requirements for factory-installed wiring do not apply to anything we are covering here or to the work of normal electricians.

Requirements Differences

The nature of the construction of a mobile home requires some deviations from the wiring of a permanent residence. From the *NEC* definition, it would appear that when mounted on a foundation, this home would cease to be a mobile home, and it is entirely possible that inspection authorities might classify it as a modular or prefabricated home, subject to the same wiring requirements and regulations as any other residence.

Due to its portability and its being mounted on a chassis and running gear, there are requirements that must be different from those for wiring in a permanent residence. This is not to imply that safety can be overlooked, especially when many mobile homes have metal siding that might become electrically energized. It might be stated that extra safety precautions should be taken in wiring a mobile home. Also, the fact that it has to be transported on wheels to a site location must not be overlooked.

Underwriters' Laboratories, Inc. (UL), has a labeling and inspection service for mobile homes that, the author is sure, would be acceptable to most inspection authorities. Where the homes are not UL approved and so labeled, the local inspection authority in many cases will want to inspect the wiring during and after completion.

Some of the differences between the wiring of a mobile home and a standard residence will be covered here. The subject will not be covered in its entirety, so reference to the current *National Electrical Code* will be necessary for specific details. Points not covered in *Article 550* of the *NEC* will be subject to the provisions of the *NEC* as covered in the wiring of a regular residence.

There is very little difference in the wiring methods used by electricians when adding wiring to a mobile home. The same methods used for adding wiring to any other residence apply. There tends to be more under-floor work in mobile homes, simply because that area is available for use. In these cases, it is important to seal the areas of the floor where cabling passes through it.

Service and Feeders

The mobile home service equipment shall be located adjacent to the mobile home, and not in or on the mobile home. The power supply to the mobile home will be feeder circuits consisting of not more than three 50-ampere mobile home supply cords or of a permanently installed circuit.

It is critical to understand that the circuit breaker panel typically found inside a mobile home is not a service panel—it is a feeder

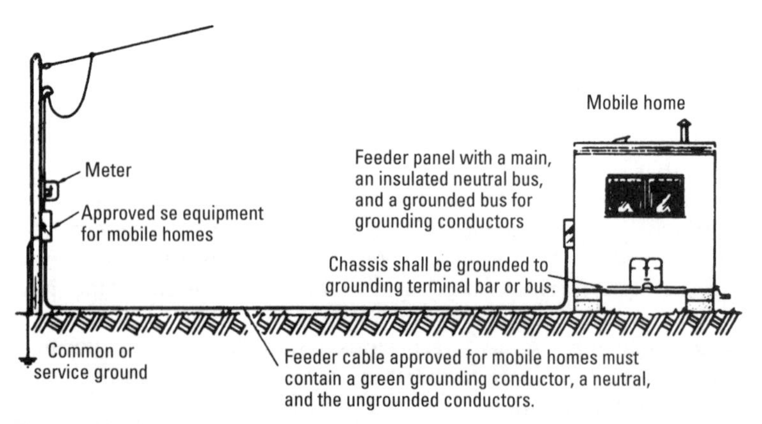

Figure 6-1 Pole-mounted service-entrance equipment for mobile home use with the feeder cable above ground.

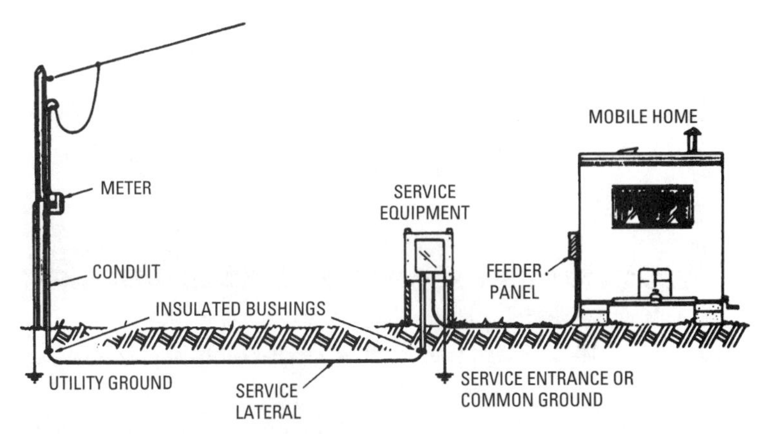

Figure 6-2 Underground service lateral for use with a mobile home.

panel. Bear in mind that this feeder panel may be the identical item used for a service panel (although the main bonding jumper should not be installed when the panel is used as a feeder panel). But even though the device may be identical, the usage is not. This is important.

Refer to Figures 6-1, 6-2, 6-3, and 6-4 for various methods of installing the service and feeders to the mobile home. Please note that the feeder panel is shown mounted on the exterior of the mobile

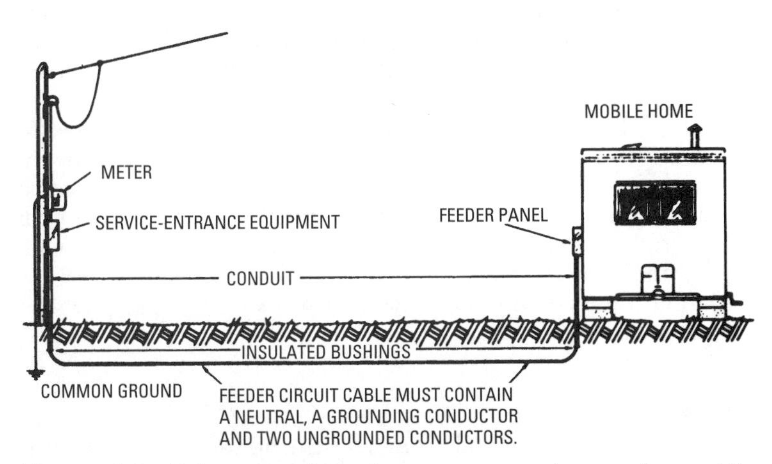

Figure 6-3 Pole-mounted service-entrance equipment for mobile home use with the feeder cable buried.

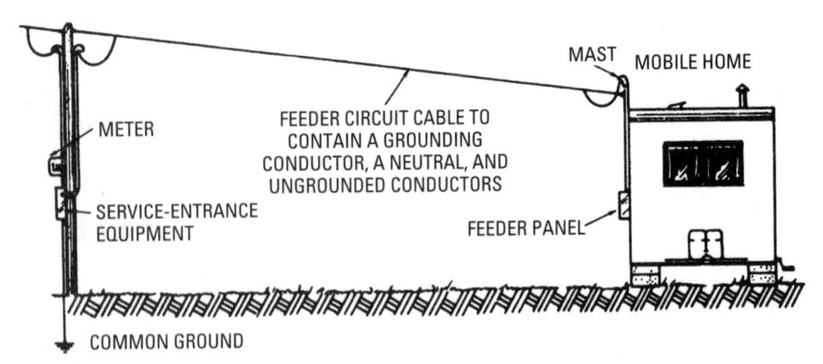

Figure 6-4 An overhead feeder cable installation to supply power to a mobile home.

home. This is merely to clarify the intent of the drawings. Usually it is mounted in the interior.

Each feeder panel fed by a cord shall not be electrically interconnected to other feeder panels fed by a cord. That is, they shall not be electrically interconnected on either the line or the load side, except that the grounding circuits and grounding means shall be electrically interconnected. Figure 6-5 illustrates the feeder-panel connections for a mobile home fed by two or fewer 50-ampere cords. Three cords were omitted in order to simplify the illustration, but the same conditions would also apply if a third supply cord were to be used.

If cord feeders are not used, but a permanent connection is made instead, such as illustrated in Figures 6-3 and 6-4, one feeder could supply and one panel would suffice. All panels are to have main disconnects with overcurrent protection.

Range and dryer frames are not permitted to be connected to the neutral as allowed for permanent structures, but are required to have a fourth equipment-grounding conductor supplied, which may be bare or green-insulated and connected to the frame, with the neutral isolated and the bonding terminal usually supplied with ranges and dryers to be discarded.

Mobile home supply cords shall not be more than 36½ feet nor less than 21 feet in length.

Protection

Because studs used in mobile home construction are often as small as 1½ inches, 16-gauge steel protection will be required to protect the

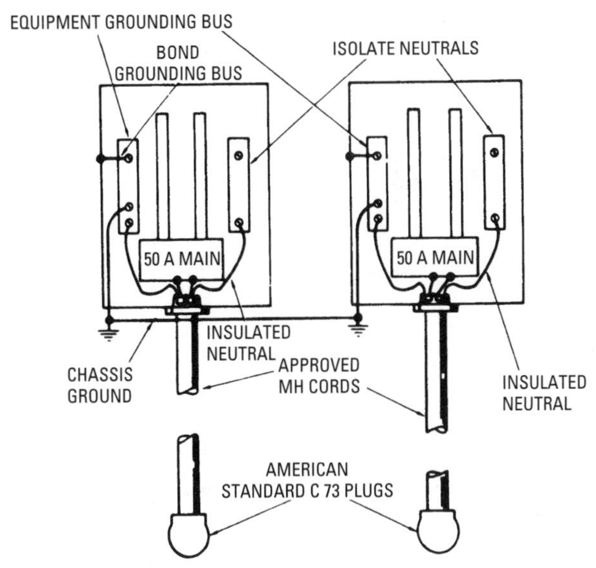

Figure 6-5 Feeder-panel connections for a mobile home fed by two 50-ampere supply cords.

wiring from nails, staples, and other fasteners. NM cable, usually used in the wiring of these homes, could be very easily penetrated by a nail or staple on these small studs. If standard 2 × 4 studs are used and are drilled in the center, this steel protection will not be required (Figure 6-6).

One very important item is the receptacle and switch box sizes. Because small joists are often used, it is hard in many cases to get boxes with the cubic content necessary to fulfill the requirements of *Article 314*. Either *Article 314* must be followed, or boxes approved for the purpose must be used.

Grounding

Section 550.16, concerning grounding, will now be quoted from the 2002 *NEC*.

> *550.16. Grounding. Grounding of both electrical and nonelectrical metal parts in a mobile home shall be through connection to a grounding bus in the mobile home distribution panel. The grounding bus shall be grounded through the green-colored insulated conductor in the supply cord or the feeder wiring to*

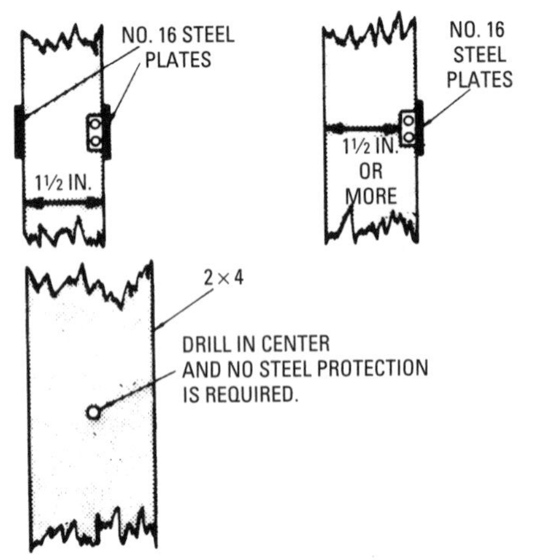

Figure 6-6 Methods of protecting mobile home wiring.

the service ground in the service-entrance equipment located adjacent to the mobile home location. Neither the frame of the mobile home nor the frame of any appliance may be connected to the neutral conductor in the mobile home.

(A) Insulated Neutral.

(1) The grounded circuit conductor (neutral) shall be insulated from the grounding conductors and from equipment enclosures and other grounded parts. The grounded (neutral) circuit terminals in the distribution panelboard and in ranges, clothes dryers, counter-mounted cooking units, and wall-mounted ovens shall be insulated from the equipment enclosure. Bonding screws, straps, or buses in the distribution panelboard or in appliances shall be removed and discarded.

(2) Connections of ranges and clothers dryers with 120/240-volt, 3-wire ratings shall be made wth 4-conductor cord and 3-pole, 4-wire, grounding-type plugs, or by Type AC cable, Type MC cable, or conductors enclosed in flexible metal conduit.

(B) Equipment Grounding Means.

(1) The green-colored insulated grounding wire in the supply cord of permanent feeder wiring shall be connected to the

grounding bus in the distribution panelboard or disconnecting means.

(2) In the electrical system, all exposed metal parts, enclosures, frames, lamp fixture canopies, etc., shall be effectively bonded to the grounding terminal or enclosure of the distribution panelboard.

(3) Cord-connected appliances, such as washing machines, clothes dryers, refrigerators, and the electrical system of gas ranges, etc., shall be grounded by means of a cord with grounding conductor and grounding-type attachment plug.

(C) Bonding of Noncurrent-Carrying Metal Parts.

(1) All exposed noncurrent-carrying metal parts that may become energized shall be effectively bonded to the grounding terminal or enclosure of the distribution panelboard. A bonding conductor shall be connected between the distribution panelboard and accessible terminal on the chassis.

(2) Grounding terminals shall be of the solderless type and approved as pressure-terminal connectors recognized for the wire size used. The bonding conductor shall be solid or stranded, insulated or bare, and shall be No. 8 copper minimum, or equal. The bonding conductor shall be routed so as not to be exposed to physical damage.

(3) Metallic gas, water, and waste pipes and metallic air-circulating ducts shall be considered bonded if they are connected to the terminal on the chassis [see *Section 550.16(C)(1)*] *by clamps, solderless connectors, or by suitable grounding-type straps.*

(4) Any metallic roof and exterior covering shall be considered bonded if (a) the metal panels overlap one another and are securely attached to the wood or metal frame parts by metallic fasteners, and (b) if the lower panel of the metallic exterior covering is secured by metallic fasteners at a cross member of the chassis by two metal straps per mobile home unit or section at opposite ends.

The bonding strap material shall be a minimum of 4 inches (102 mm) in width of material equivalent to the skin or a material of equal or better electrical conductivity. The straps shall be fastened with paint-penetrating fittings, such as screws and starwashers or equivalent.

550.17. Testing.

(A) Dielectric Strength Test. The wiring of each mobile home shall be subjected to a 1-minute, 900-volt, dielectric strength test (with all switches closed) between live parts (including neutral) and the mobile home ground. Alternatively, the test shall be permitted to be performed at 1,080 volts for 1 second. This test shall be performed after branch circuits are complete and after fixtures or appliances are installed.

Exception: Listed fixtures or appliances shall not be required to withstand the dielectric strength test.

It is highly recommended that this test be run at the completion of the wiring of each mobile home, and it is further recommended that it also be run when the mobile home is delivered to the site and also after each subsequent move. Notice that testing at each site is not a *Code* requirement, but it would certainly be a step toward better safety.

Each receptacle should be checked for proper polarity and equipment ground continuity. In checking mobile wiring, it was found that the majority of rejects were due to improper grounding or polarities. Rechecks of inspections are time-consuming and expensive to both the manufacturer and the inspection authority.

Chapter 7

Cable TV, Broadband, Telephone, and Security

Because of the increasingly widespread use of communications and security systems over the past two decades, electricians have been called upon to install many different types of wiring systems in homes. Although all these systems are composed of wires carrying electricity, the similarity to conventional wiring more or less ends there. Because of this distinction, this chapter will cover these systems:

1. Cable TV wiring
2. Telephone wiring
3. Broadband and home networks
4. Security wiring

Cable TV Wiring

Wiring for cable TV systems is covered by *Article 820* of the *NEC*. This article is specifically written for community antenna TV (CATV) systems but is applicable to common cable TV wiring. *Part A* of *Article 820* covers general requirements, which do not apply directly to the installer. *Part I* covers outdoor cables and the protection and grounding of cables. These requirements apply to the installers for the cable TV supplier rather than to the electrician who installs the wiring in the house.

For wiring within the residence, we refer to *Part V* of *Article 820 (Cables Within Buildings)*.

Section 820.50 specifies that the coaxial cables used for cable TV systems must be suitable for the purpose and resistant to the spread of fire. Usually Type RG59U cable is used.

Section 820.52 covers the installation of the coaxial cables. Generally, they may not be installed in the same raceway as power conductors or with Class 1 circuits. They may be run in the same raceway with Class 2 and 3 circuits. See *Article 725* for explanation of Class 1, 2, and 3 circuits. When installed as open wiring (as is common in homes), the coaxial cables must be kept at least 2 inches away from power or Class 1 circuits. When installed in plenums, these cables must be Type CATVP or meet the requirements of *Section 300.22*.

Although different cable TV suppliers may have different requirements (and you should verify the requirements *before* you begin any installation), the general installation procedure is as follows:

Usually the best method of installation is to bring a lead-in cable from the exterior of the home to a central accessible location, usually in an attic. From there, runs to individual outlets are brought out, like the tentacles of an octopus. The cables should be stapled to the framing members in the same manner as Type NM power cable (see Chapter 4 of this book). Care should be taken not to drive the staples in too far, because crimping the cable will harm its operation.

By arranging for all the cables to originate at one central location, the cable TV supplier can easily install the various amplifiers and splitters. This also makes the system far easier to service.

At the outlets, a plastic outlet box is attached to the wall stud, just as for a receptacle outlet, and the coaxial cable pulled into the box. Generally, it is recommended to leave 5 feet of extra cable at the outlets to facilitate connection to the television set. Remember, however, that this cable should be protected from damage until the construction is complete.

After the walls are finished, a trim plate having one hole is installed over the outlet, and the coaxial cable threaded through the hole. Typically, the installer's responsibility stops at this point.

Telephone Wiring

Generally, a four-wire telephone cable is looped from outlet to outlet, beginning at the location on the exterior of the house where the utility will connect the cable to its system, and ending at the final telephone outlet.

Like coaxial cable used for cable TV, the telephone cable should be installed in the same manner as Type NM power cable (see Chapter 4 of this book), and it should be terminated in plastic boxes, just as for cable TV outlets. Note, however, that wall phones and phones mounted over desks will need to be at special heights. Verify the heights of these outlets before installation.

Sometimes installers are required to put a trim plate over the box, and sometimes they may be required to install the jack. Again, verify these details with the local service supplier prior to installation.

A four-wire cable will handle two telephone lines (two separate phone numbers); if more lines are required, separate runs will be required.

Article 800 of the *NEC* covers communication circuits such as telephone systems. These circuits must be separated from power circuits and must be grounded. In addition, all such circuits that run outdoors (even if only partially) must be provided with circuit protectors (surge or voltage suppressors).

The sections below list the requirements for installation of communication circuits. Only one section applies to telephone wiring inside a house: *Interior Communications Conductors*, on page 139. Nonetheless, we are also including sections that deal with circuits outside a house. These would apply to conductors run to a garage or outbuilding and are good to have as a reference.

Conductors Entering Buildings

If communications and power conductors are supported by the same pole or run parallel in span, the following conditions must be met:

1. Wherever possible, communications conductors should be located below power conductors.
2. Communications conductors cannot be connected to crossarms.
3. Power service drops must be separated from communications service drops by at least 12 inches.

Above roofs, communications conductors must have the following clearances:

1. Flat roofs: 8 feet
2. Garages and other auxiliary buildings: none required
3. Overhangs, where no more than 4 feet of communications cable will run over the area: 18 inches
4. Where the roof slope is at least 4 inches in rise for every 12 inches horizontally: 3 feet

Underground communications conductors must be separated from power conductors in manholes or hand holes by brick, concrete, or tile partitions.

Communications conductors should be kept at least 6 feet away from lightning protection system conductors.

Circuit Protection

Protectors are surge arresters designed for the specific requirements of communications circuits. They are required for all aerial circuits not confined within a city block. They must also be installed on all circuits within a block that could accidentally contact power circuits over 300 V to ground. They must also be listed for the type of installation.

Other requirements are as follows:

1. Metal sheaths of any communications cables must be grounded or interrupted with an insulating joint as close as practicable to the point where they enter any building (such point of entrance being the place where the communications cable emerges through an exterior wall or concrete floor slab or from a grounded rigid or intermediate metal conduit).
2. Grounding conductors for communications circuits must be copper or some other corrosion-resistant material and must have insulation suitable for the area in which the conductors are installed.
3. Communications grounding conductors may be no smaller than No. 14.
4. The grounding conductor must be run as directly as possible to the grounding electrode and must be protected if necessary.
5. If the grounding conductor is protected by metal raceway, the raceway must be bonded to the grounding conductor on both ends.

Grounding electrodes for communications ground may be any of the following:

1. The grounding electrode of an electrical power system.
2. A grounded, interior, metal piping system. (Avoid gas piping systems for obvious reasons.)
3. Metal power service raceway.
4. Power service equipment enclosures.
5. A separate grounding electrode.

If the building being served has no grounding electrode system, the following can be used as a grounding electrode:

1. Any acceptable power system grounding electrode (see *Section 250.52* of the *NEC*).
2. A grounded metal structure.
3. A ground rod or pipe at least 5 feet long and 1/2 inch in diameter. This rod should be driven into damp (if possible) earth and should be kept separate from any lightning protection system grounds or conductors.

Connections to grounding electrodes must be made with the *NEC* approved means. If the power and communications systems use separate grounding electrodes, they must be bonded together with a No. 6 copper conductor. Other electrodes may be bonded also. This is not required for mobile homes.

For mobile homes, if there is no service equipment or disconnect within 30 feet of the mobile home wall, the communications circuit must have its own grounding electrode. In this case, or if the mobile home is connected with cord and plug, the communications circuit protector must be bonded to the mobile home frame or grounding terminal with a copper conductor no smaller than No. 12.

Interior Communications Conductors

Communications conductors must be kept at least 2 inches away from power or Class 1 conductors, unless they are permanently separated from each other or unless the power or Class 1 conductors are enclosed in one of the following:

1. Raceway
2. Type AC, MC, UF, NM, or NM cable, or metal-sheathed cable

Communications cables are allowed in the same raceway, box, or cable with any of the following:

1. Class 2 and 3 remote-control, signaling, and power-limited circuits
2. Power-limited fire-protective signaling systems
3. Conductive or nonconductive optical fiber cables
4. Community antenna television and radio distribution systems

A few additional notes also apply:

- Communications conductors cannot be in the same raceway or fitting with power or Class 1 circuits.
- Communications conductors cannot be supported by a raceway unless the raceway runs directly to the piece of equipment the communications circuit serves.
- Openings through fire-resistant floors, walls, and so on, must be sealed with an appropriate fire-stopping material.
- Any communications cables used in plenums or environmental air-handling spaces must be listed for such use.
- Communications and multipurpose cables can be installed in cable trays.

- Any communications cables used in risers must be listed for such use.

Cable substitution types are shown in *Table 800.53* of the *NEC*.

Telephone Connections

The most common and simplest type of communication installation is the single-line telephone. The typical telephone cable contains four wires, colored green, red, black, and yellow. A one-line telephone requires only two wires to operate. In almost all circumstances, green and red are the two conductors used. In a common four-wire modular connector, the red and green conductors are found in the inside positions, with the yellow and black wires on the outer positions.

As long as the two center conductors of the jack (again, always green and red) are connected to live phone lines, the telephone should operate.

Two-line phones generally use the same four-wire cables and jacks. In the case of two-line phones, however, the inside two wires (red and green) carry line 1, and the outside two wires (black and yellow) carry line 2.

Broadband and Home Networks

Broadband Internet connections are actually very easy to obtain in most places, and prewiring a home for them has become almost unnecessary. The most common broadband connects are either DSL connections, which come through a standard telephone line, or cable modem connections, which come through a cable television system. In both cases, all that is required is a small box and a few connecting cables. The only prewiring required is to make sure that there is a telephone cable or coaxial television cable where needed.

To take broadband service from the point of entrance to other locations in the home, however, requires additional work. There are two choices:

1. Distribution through the house via cable
2. Distribution through the house via wireless

Before we go through some of the basics, it is important to define a few terms:

A *repeater* is a device that receives and then immediately retransmits each bit. It has no memory and does not depend on any particular protocol. It duplicates everything.

A *bridge* receives an entire message into memory, analyzes it, and then retransmits it. If the message was damaged by a collision or noise, it is discarded. If the bridge knows that the message was being sent between two stations on the same cable, it discards it. Otherwise, the message is queued up and will be retransmitted on another cable. The bridge has no address. Its actions are transparent to the client and server workstations.

A *router* acts as an agent to receive and forward messages. The router has an address and is known to the client or server machines. Typically, machines directly send messages to each other when they are on the same cable, and they send the router messages addressed to another zone, department, or sub-network.

10Base5, 10BaseT, 10Broad36, and the like are the IEEE names for the different physical types of local area network often called Ethernet. The "10" stands for a signaling speed, of 10MHz. Base means baseband, and Broad means broadband. The last section indicates the cable type, where T means twisted pair and F (as in 10BaseF) means fiber.

An *access point* is a device that transports data between a wireless network and a wired network.

A *wireless node* is a user computer with a wireless network interface card (adapter).

Home Networks

Home networks are essentially light versions of commercial computer networks. They require a broadband connection to the Internet and a central panel. Depending on the complexity of the network, there may be connections to telephone and cable TV systems as well. From the panel, twisted-pair cables are run through the house, terminating in wall outlets in the same way as telephone cables. At the outlets, special plugs are installed, and short cords (called *patch cords*) connect computers to the outlets. If telephone and cable TV service is to be provided by the system, the process is the same, except that the cables and plugs vary.

ANSI/TIA/EIA-570-A Residential Telecommunications Cabling Standard covers home networking. This standard is derived from the usual EIA/TIA 568 standard for commercial structured cabling systems. Refer to Figures 7-1 and 7-2.

The *NEC* requirements for home networking are the same as for interior telephone conductors, which were covered earlier in this chapter. See *Article 800* of the *NEC* for telephone-type cables, and *Article 820* for coaxial cables, such as are used for cable TV circuits.

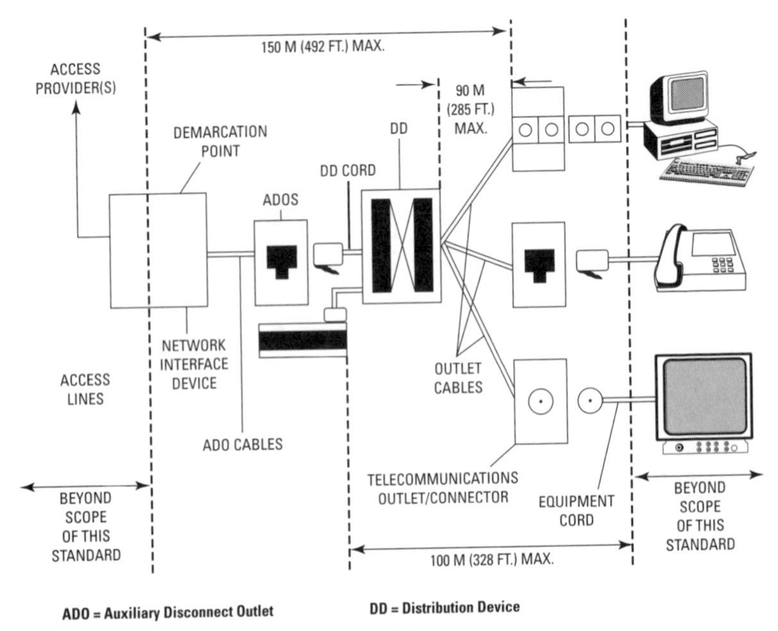

Figure 7-1 Typical home network layout.

Wireless

Wireless networks are a very effective and affordable method of connecting multiple home computers to the Internet. The signal transmission capacities and distance limits are not as great with wireless systems as they are for hard-wired systems, but they are generally more than sufficient.

When a wireless system is to be installed in a home, the main wireless panel will be installed next to or near the main telecommunications entrance to the home and will connect directly to the broadband device—a cable modem or DSL hub. From there, the signal will be distributed electromagnetically, with each connected device being fitted with a receiver and connecting hardware.

The simplest wireless network configuration is an independent (or peer-to-peer) network that connects a set of personal computers (PCs) with wireless adapters. Any time two or more wireless adapters are within range of each other, they can set up an

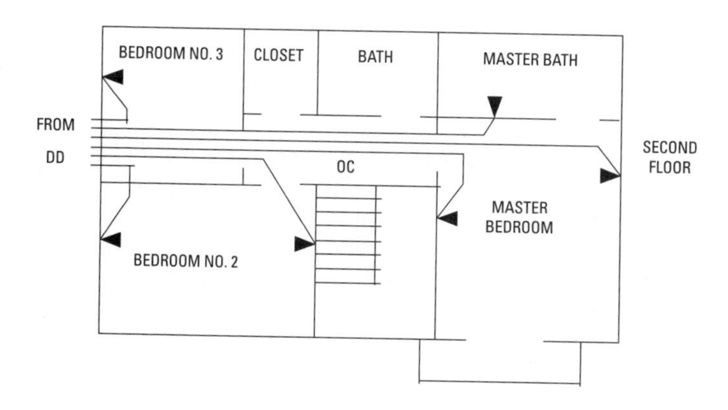

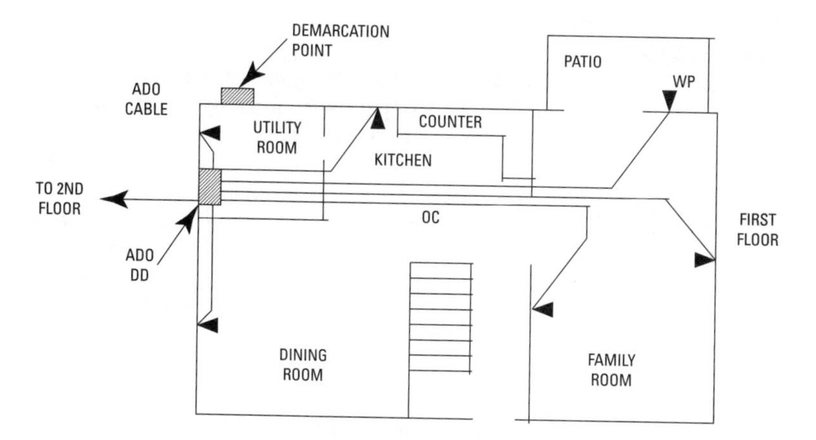

NOTE – Some code bodies limit placing telecommunications outlet/connectors in bathrooms.

Legend:

▲ – Telecommunications Outlet/Connector
ADO – Auxiliary Disconnect Outlet
DD – Distribution Device
OC – Outlet Cable
WP – Waterproof Outlet Box

Figure 7-2 Typical cabling system for a single residential unit.

independent network. These on-demand networks typically require no administration or preconfiguration. Access points can extend the range of independent networks by acting as a repeater, effectively doubling the distance between wireless PCs.

The distance over which radio signals can be used to communicate is a function of product design (including transmitted power and receiver design) and the propagation path, especially in indoor environments. Interactions with typical building objects, such as walls, metal, and even people, can affect how the magnetic energy propagates, and thus the range and coverage that a system achieves. Radio waves can penetrate many indoor walls and surfaces. The range (or radius of coverage) for typical wireless network systems varies from under 100 feet to more than 500 feet. Coverage can be extended.

Network Speed

As with wired networks, the actual rate of signal transfer (also called throughput) in wireless LANs is dependent on the product and the setup. Factors that affect throughput include the number of users, the type of system used, and bottlenecks on the wired portions of the system. Typical data rates range from 1 to 10 megabits per second (Mbps). Users of traditional Ethernet networks generally experience little difference in performance when using a wireless network. Wireless networks provide throughput sufficient for the most common office applications, including e-mail, shared peripherals, and multiuser databases and applications.

Security

One of the root technologies for wireless networks is military communications. Because of this, security has long been designed into most systems. These security provisions frequently make wireless networks more secure than most wired networks. Complex encryption techniques make it impossible for all but the most sophisticated to gain unauthorized access to network traffic. In general, individual nodes must be security-enabled before they are allowed to participate in network traffic. Nonetheless, unless security is taken seriously by the user, it can be an issue. And it must be said that Microsoft products are especially susceptible to security breaches.

The unlicensed nature of radio-based wireless networks means that other products that transmit energy in the same frequency spectrum can cause interference with a wireless network. Microwave ovens are a potential concern, but most manufacturers design their products to account for microwave interference.

Security Systems

All security systems, regardless of how simple or complex, can be divided into three basic parts:

- **Sensors.** These are the devices that sense or respond to certain conditions in and around the protected area. (There are many types of sensors for many different applications.)
- **Controls.** The brains of the system are these controls that respond to the input from the sensors according to the desires of whoever set the system up.
- **Signaling devices.** These devices give out some type of signal (most typically a siren or buzzer) when an alarm condition is reached.

The critical factors in applying these devices to a home are that:

- All devices must be compatible.
- All the devices must be connected properly.
- The system must be properly designed to cover all vulnerable areas of the home and to provide effective responses.

Wired or Wireless

The primary choice in security systems is between wired systems, which have cables running to every device, and wireless systems, which transfer signals via radio waves (but which require regular battery changes).

Wired components are less expensive than wireless systems, but wired systems involve higher installation expenses.

The benefit gained by going wireless in that the system is less labor-intensive to install. In particular, it is the cost of installing cables that makes a wired system expensive, especially in existing homes. By sending signals via radio waves, this cost is completely avoided.

Yet, although a good deal of money is saved by avoiding the costly installation of wires, the wireless devices themselves cost quite a bit more than do the corresponding conventional devices. The reason is that each device must have two parts: a sensor and a transmitter. Additionally, they must have some type of built-in power source. Because of these requirements, the wireless devices are not only more expensive but also physically larger than the standard components.

Wireless window switches are made to be wired directly into a transmitter, which then sends a signal to the receiver when necessary. Most systems have the receiver built right into the control unit, although some use a separate receiver. Sometimes separate receivers are required for outdoor or special transmitters. A typical transmitter has a range of about 200 feet. If the transmitter is farther from the main receiver than this distance, a separate receiver must be installed within the transmitter's range.

Security Installations

Security wiring is covered by *Article 725* of the *NEC*, and security circuits are generally classified as Class 2 or 3. The installation requirements are similar to those for coaxial cable.

When installing security circuits, remember that there are two basic types of security circuits:

- Detection circuits, which run to the various devices that are used to detect an intrusion
- Signaling circuits, which run to the various devices (horns, lights, sirens, and the like) that signal an intrusion

These circuits cannot be joined at any time. Each type must have its own path to the control panel. For most systems, a small resistor will have to be installed at the end of each circuit. These are called end-of-the-line (EOL) resistors. These are required for monitoring the system.

For the installation of security wiring, it is critical that you follow the system manufacturer's instructions to the letter. To fail to do so will almost always result in problems. These are sensitive systems and must be installed exactly as designed. If you encounter a problem during the installation, you should call the supplier of the system rather than trying to figure out the easiest answer. Security wiring is installed in the same manner as telephone wiring and is terminated in plastic boxes in the same manner. Again, verify the mounting heights of all devices prior to roughing them in.

If you are required to install the various security devices and/or make the final connections to the control panel, make sure that the manufacturer's instructions are followed exactly. Again, don't try to find an easy answer to problems that may occur; call an engineer who is thoroughly familiar with the system.

Chapter 8

How to Wire a House

In this chapter, we will go through the actual process of wiring a house. In previous chapters we focused on the rules of such installations. In this chapter we will focus on both the requirements and the process.

Figure 8-1 shows a typical layout for wiring the home shown in Figure 3-3, part A, on page 52. These drawings will be used in this summary as well as the load calculations used in Chapter 3 of this book. Not only will the minimum *Code* requirements be used; there will also be suggestions for a more adequate wiring job and additions in the future for additional usage.

The Process

We begin this chapter by covering the steps required to wire a house. The process generally consists of the following:

1. Planning
2. Slab work
3. Rough-in
4. Trimming

These are actually quite distinct processes, although there can be some overlap and extension. For example, the trimming stage can sometimes extend into a final closeout of the job, where a final few corrections must be made at the request of the electrical inspector.

Planning

The first steps in wiring a house are to design the wiring system and to obtain the necessary permits. Generally, these steps require submitting blueprints and a load calculation, discussing the project with an inspector, and paying a fee. The procedure entails an initial visit with the local electrical inspector, then subsequent visits and payments to complete the process. It will also require coordination with the general building permit and building contractor.

The next step is to coordinate with the other people involved in the construction process. Walls must be up before wiring is installed in or on them, floors must be in place, and so on. Scheduling and coordination on a construction site are extremely important. If done well, the project can move forward very nicely; if not, every day is a series of problems, obstacles, and frustrations. Get realistic

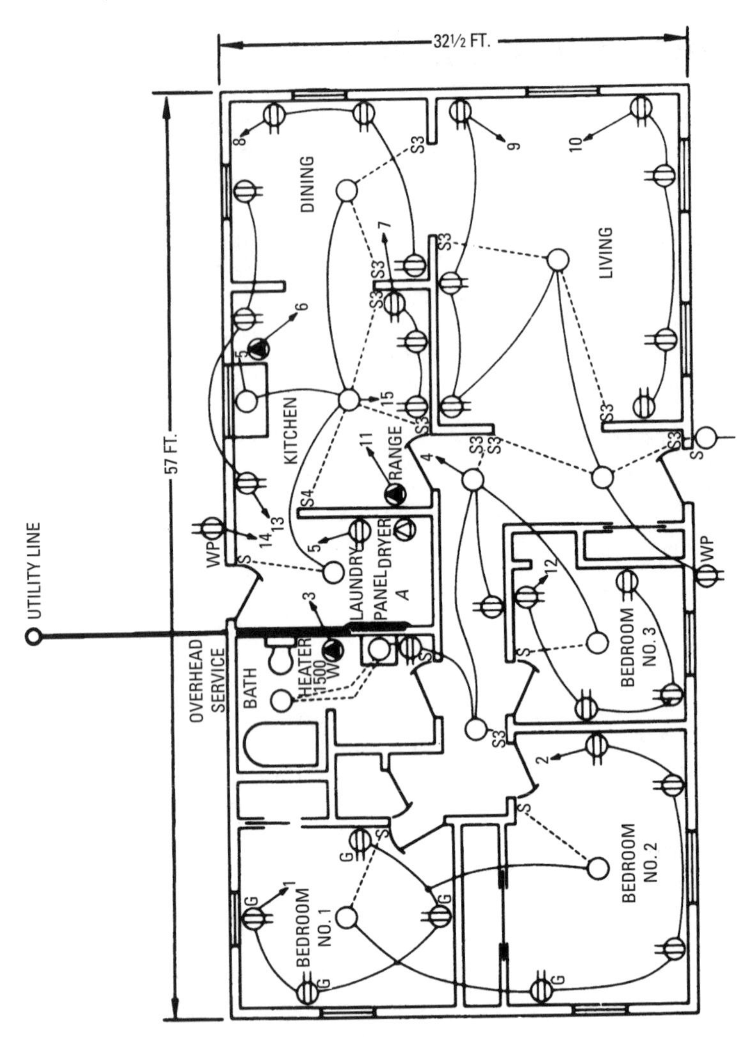

Figure 8-1 Service-entrance location.

schedules and keep them updated. This is an ongoing and sometimes difficult process.

Next, you must plan material deliveries. The items you need must be on the job when you need them. If not, you'll waste large amounts of time running to and from suppliers. You should either order only the materials you'll need for each portion of the project or else have material storage facilities on the site. This will require you to prepare accurate material lists and to place these orders a couple of days ahead of the delivery time.

In addition to materials, you will need tools and workers. Even if you are wiring a house yourself, you will need competent help for parts of the job. If the necessary people are not there when you need them, the project will become frustrating, and the quality of the job is likely to suffer. In order to find such workers, any of several avenues are available: You can find an acquaintance who is, or knows, a professional electrician; you can place a help wanted ad; or you can hire a local electrician.

Finally, you will have to coordinate inspections. This will require phone calls to the local inspector a few days ahead of time. Without inspections the house may not be legally occupied. But equally important is that you must keep the local inspector reasonably happy. It is eminently in your interest to work with the inspector, rather than to become at odds.

The Slab

Running wiring underneath the floor slab of a house is not uncommon. Usually, underslab conduits run between wall outlets, telephone of cable TV runs, or for an underground service. Any type of circuit can be run underslab in conduit, but the difficulties in placement usually restrict such runs to outlets located on exterior walls. Interior walls are much harder to locate than exterior walls, which will be easily found during the construction process. If done, PVC conduit must be installed after the ground is leveled and fully prepared but before the slab is poured.

The first issue is that the conduit is placed correctly. You must determine exactly where the walls will be and place your conduits so that they will come up either inside of or next to the wall (as the case may be). Attach the vertical runs of conduit to any wall that may be available, or drive a stake into the ground, and attach the conduit to that.

The horizontal runs of conduit go under the concrete. A small shovel can be used to dig a small trench. Then place the conduit and re-cover it.

Again, correct placement of the vertical runs cannot be overstressed. If you miss a wall, the conduit will have to be cut off and abandoned.

The Rough-In

Roughing-in is the process of installing wiring or raceways inside of walls before they are finished. This is done about midway through the construction of a house, as soon as the walls are framed-in. The following is the rough-in procedure of a long-time residential electrical contractor:

1. Unload all materials and tools that will be used that day. Place all materials and tools in an accessible and central location.
2. Run an extension cord to the power source (either a temporary electric service, an adjoining home, or a mobile generator).
3. Fill and hang the required cable reels. (The contractor wired houses with Type NM cable, which was obtained on reels. The reels were mounted at central locations several feet above the floor, with conduit as an axle so that the reel could unroll easily.)
4. Mount all of the electrical boxes. Set receptacles at 14 inches on center (OC); switches at 48 inches OC; outlets above counter-tops at 42 inches OC. Use pancake boxes screwed to the bottom of ceiling joists for ceiling fans and light fixtures with only one cable running to them. Use round boxes for other light fixtures.
5. Mount plaster rings for low-voltage outlets (cable TV and telephone).
6. Drill all holes and notch any framing members as required. Use a 1-inch auger bit with the right-angle drill for drilling studs and plates. Drill several large holes above load center for home runs, and more holes if required.
7. Install cable *exactly as shown on plans.* As each area is completed, highlight plans. Install branch-circuit cable first, then TV and phone wire, followed by power circuits (dryer, range, A/C units, heater, and so forth). Leave 8 to 10 inches of cable coming out of all boxes. Run any special circuits last (post lights, well pumps, and so forth). Mark all cables as required during the installation.
8. Install the service panel. Pull the required knockouts and install the connectors before mounting the panel. Use a large terminal adaptor and locknut (3 in.) in the top knockout for bringing branch circuits into the panel (where approved). Run service cable.

9. Bring all cables into the panel. Be sure that all are marked. Use a connector for service cable. Cover the panel front with cardboard. This keeps paint and drywall mud off of your bus bars during the processes of finishing the walls.
10. Mount and wire the meter base.
11. Drive the ground rod and run the grounding electrode conductor. Run grounding conductor from the panel to a water pipe.
12. Install the service riser for overhead service or the drop to underground service, if required. Install the lightning arrestor if required.
13. Splice all branch-circuit cables.
14. Double-check the entire job to be sure that everything is complete. Load all materials and tools back into the truck. Thoroughly clean the job site.

Trimming

Trimming-out a job involves wiring the actual receptacles, switches and light fixtures, mounting all finish plates, and making all connections throughout the house—in short, finishing the job.

Obviously, trimming must be done carefully and competently. In this task, it is also critical to have all of the required materials available. Missing items will make trimming long and frustrating.

A final step that usually occurs at the same time as trimming is final testing. Every outlet in the house should be tested, and every electrically-operated device checked for proper operation. Although this step needn't take a great deal of time, it must be done meticulously.

Service-Entrance Location

The first important consideration in planning the wiring of a house is where to locate the service-entrance equipment. There are several items that will enter into making a decision:

1. The *NEC*, *Section 230.70(A)*, states that the disconnecting means shall be located at a readily accessible point nearest the entrance of the conductors, either inside or outside the building or structure. *Sections 230.70(A)(1)* and *230.70(A)(2)* add that sufficient access and working space shall be provided about the disconnecting means.

This requirement is short but very meaningful. What is the closest point of entrance? Many authorities consider 15 feet as a maximum. However, this varies. Why the problem? The service-entrance conductors are without overcurrent protection until they arrive at the main disconnect; thus, the length of unprotected conductors must be kept to a minimum for fire and safety protection.

2. A utility must supply service to the residence; therefore, its requirements must be considered as to where power lines are located and related issues. Thus, one must consult with the utility as to location. If the house is supplied by a service lateral (underground service), the problem eases to quite a degree. The service equipment is at a very good point if the utility can serve from the point shown in Figure 8-1.

Take a look at Figure 8-2. Here the utility line is at the end of the home, which leaves us with several alternatives:

1. The main disconnect and overcurrent device may be located at the end of the house, on the outside in a raintight enclosure and a feeder circuit (three-wire with equipment-grounding conductor) run to the branch circuit panel *A*.
2. A service lateral may be run, as shown by the dashed lines in Figure 8-2.

These, of course, are not all of the answers. Each case must stand on its own merits, taking into consideration all of the following points, which *shall be* met:

a. The disconnecting means shall be located "*nearest to the entrance of the conductors, either inside or outside the building.*"
b. The disconnecting means must be "*readily accessible.*" Do not place the service equipment over a washer, dryer, or similar large appliance; near combustibles, such as in a closet, in a bathroom or bedroom; or too high.
c. The disconnecting means requires easy access in the case of emergencies.

One thing should be uppermost in the mind of anyone wiring a home, and that is that most people spend the greater part of their working life making payments on a home, so safety and adequate wiring should be in the mind of every electrician in planning the wiring and in the workmanship used.

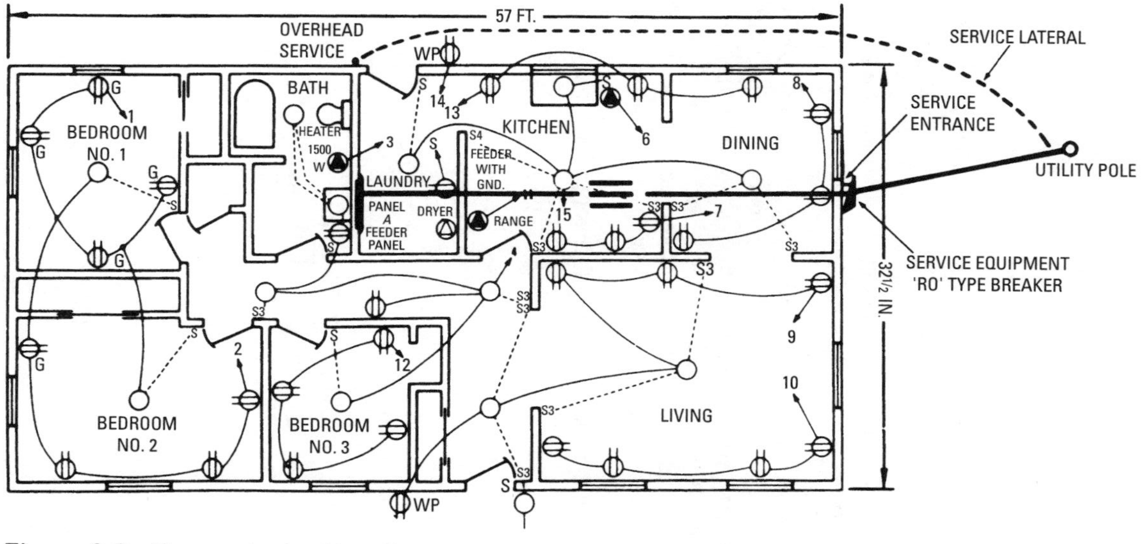

Figure 8-2 Two methods of installing service when the utility pole location does not fit equipment location.

Service Size

What size shall the service be? In Chapter 2, we stated that the service for a single-family residence should be 100 amperes, but remember that this is the minimum requirement. Going significantly beyond the minimum is strongly recommended. With the added usage of electricity, we will no doubt overload the 100-ampere service in the future. Plan now to increase the size of the service to 150 or 200 amperes. The added cost is not large, and the cost of increasing the size of the service later would be much greater. To be competitive on bids, there is nothing to keep you from making a bid on the 100-ampere service and then making an alternate bid for 150- or 200-ampere service and doing a little selling.

In Calculation No. 1 of Appendix A, we find that six 15-ampere circuits would be required (minimum) for the general lighting load plus a minimum of two 20-ampere circuits for the small-appliance load and one 20-ampere circuit for the laundry circuit; a 35-ampere (two-pole) circuit for the range and a 40-ampere (two-pole) circuit for the dryer are also figured. These requirements are summarized in *Table 8-1*. There should also be a few spares for future additions.

Table 8-1 Circuit Requirements for House

6	15-ampere circuits
3	20-ampere circuits
2	35-ampere circuits (figuring 2 poles)
2	40-ampere circuits (figuring 2 poles)
13 circuits minimum	

Table 8-2 Standard Branch-Circuit Panels

Number of Single Poles
12
16
24
30
40
42

Standard branch-circuit panels come in the sizes listed in *Table 8-2*. The minimum size to be used would be a 16-pole panel. When purchasing the panel, you need not purchase it with breakers to fill all spaces. Look to the future. The base price of the enclosure is a very small part of the total wiring cost, but an addition later will be very costly. Boost your sights and purchase a larger enclosure and thus plan for the future.

The panel readily fits between studs spaced at 16 inches on center. Secure it solidly and bring it out flush with the finished wall surface. It is easy to install all of the cables or raceways during the rough-in, because the walls are open. After the wall is finished, it will be hard to fish in additional circuits. Install a $1\frac{1}{4}$-inch or $1\frac{1}{2}$-inch conduit or EMT into the attic area and also into the basement area to facilitate installing any additional circuits that will most certainly be added in the future (Figure 8-3).

It is sometimes impossible to install the conduit into the basement area. In this case, drill the future holes through the plate and other framing, and install lengths of single insulated conductors, to be used as fish wires for future circuits.

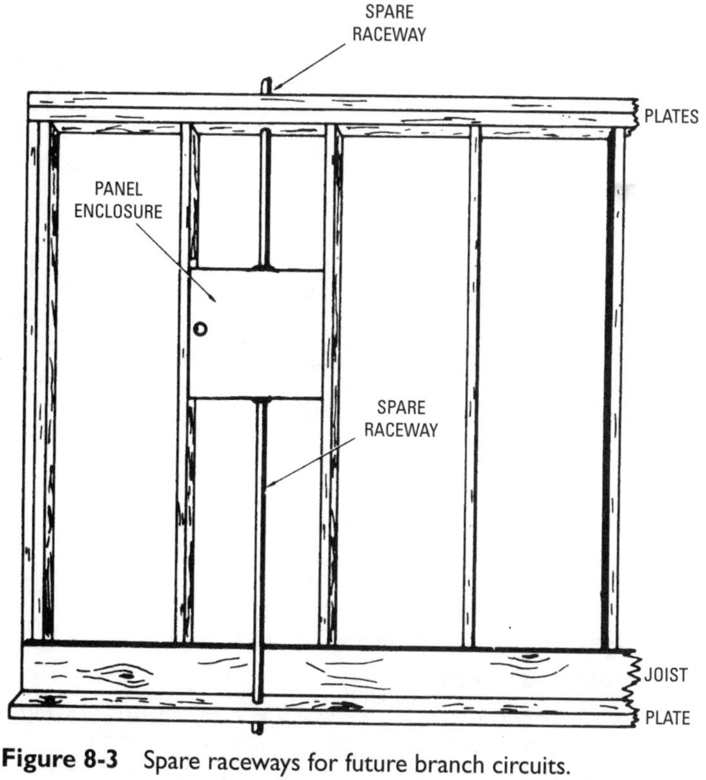

Figure 8-3 Spare raceways for future branch circuits.

It is amazing how often the inspector hears of a homeowner who has had a house wired only to the minimum *Code* requirements. The homeowner tells of additional circuits to be added but finds it practically impossible to make these additions to meet the *Code* requirements.

Bonding and grounding of services have the highest priority. Grounding bars for equipment-grounding conductors are available for installation in panels, or the enclosures may be purchased with grounding bars installed (Figure 8-4).

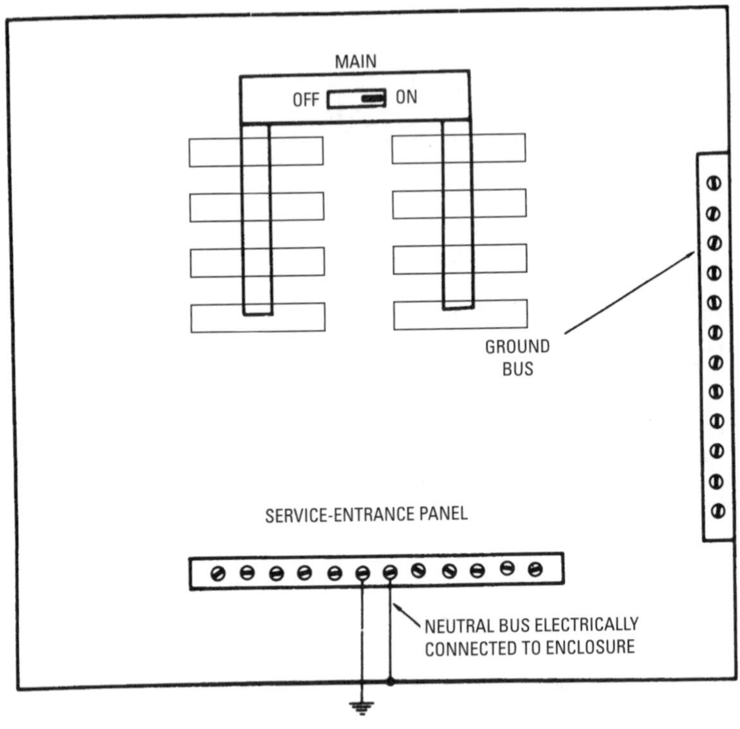

Figure 8-4 Equipment-grounding conductor in a service equipment panel.

Feeder Panels and Branch-Circuit Overcurrent Devices

The difference between *feeder panels* and *service-entrance panels*, with branch-circuit overcurrent devices in the service-entrance equipment, should be covered. A feeder is the circuit conductors between the service equipment (or the generator switchboard or an isolated plant) and the branch-circuit overcurrent device.

Most dwellings have the branch-circuit overcurrent devices in the service equipment enclosure, as would be the case in Figure 8-1. In Figure 8-2 the service equipment is outdoors at the end of the house, so the conductors to the feeder panel will be feeders and consist of two phase conductors, one neutral conductor, and one green or bare equipment-grounding conductor, sized according to *Table 250.122* of the *NEC* (reproduced here as *Table 8-3*).

Table 8-3 Minimum Size Equipment Grounding Conductors for Grounding Raceway and Equipment (in the *NEC*, *Table 250.122*)

Rating or Setting of Automatic Overcurrent Device in Circuit Ahead of Equipment, Conduit, etc., Not Exceeding (Amperes)	*Size*	
	Copper Wire No.	*Aluminum Wire No.**
15	14	12
20	12	10
30	10	8
40	10	8
60	10	8
100	8	6
200	6	4
400	3	1
600	1	2/0
800	0	3/0
1000	2/0	4/0
1200	3/0	250 kcmil
1600	4/0	350 kcmil
2000	250 kcmil	400 kcmil
2500	350 kcmil	600 kcmil
3000	400 kcmil	600 kcmil
4000	500 kcmil	800 kcmil
5000	700 kcmil	1200 kcmil
6000	800 kcmil	1200 kcmil

**See installation restrictions in* Section 250-92(A).

Another instance where we could find a feeder is in a large home with the service equipment and branch-circuit breakers in one enclosure but using a large-ampacity breaker to feed a second branch-circuit panel somewhere else about the house.

Your attention is called to Figure 8-4, which shows the neutral bus electrically connected to the enclosure. This is required in service-entrance panels. On feeder panels, on the other hand, we isolate the neutral bus from the enclosure; the green or bare equipment-grounding conductor of the feeder, just mentioned, is grounded to the feeder panel enclosure, and all branch-circuit equipment-grounding conductors are brought to a grounding strap used for this purpose (Figure 8-5).

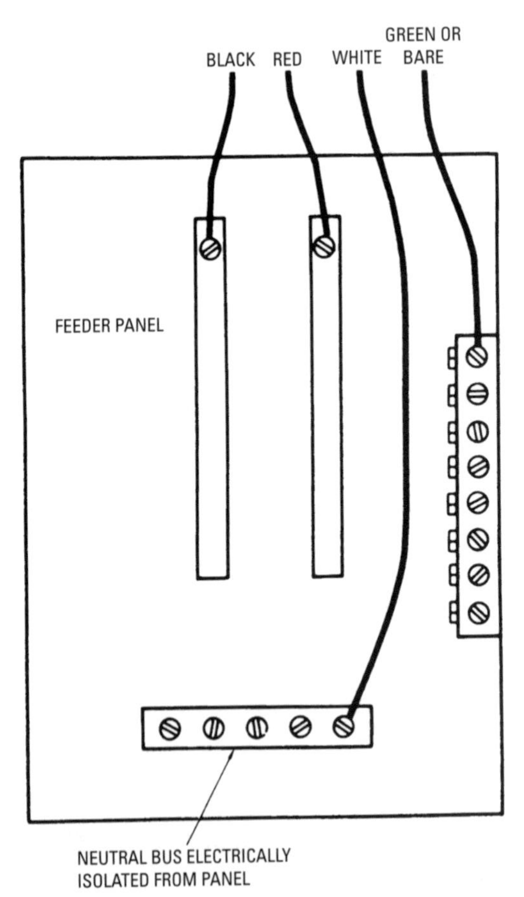

Figure 8-5 Feeder panel connections.

Number of Outlets Per Circuit

This is a controversial subject (contractors argue continually over the proper number of outlets to connect to a circuit), but it need not be so. Let us look at what the *NEC* has to say about it.

The note with *Table 220.3(A)* tells us about receptacles in single and multifamily dwellings and further tells us to refer to *Section 220.3(B)(10)*. *Article 100* tells us, *A receptacle is a contact device installed at the outlet for the connection of a single attachment plug.* It defines a multiple receptacle as a *single device containing two or more receptacles.* The *Code* defines an *outlet* as a point on the wiring system at which current is taken to supply utilization equipment. This definition includes not only receptacles, but also openings for such devices as luminaires or smoke detectors.

From these definitions, when we figure outlets for small appliance circuits, we must consider a duplex receptacle as two outlets and use 180 volt-amperes per outlet, which may be broken down into amperes:

$$180\ \text{VA}/120\ \text{V} = 1.5\ \text{amperes}$$

A small appliance circuit may have 13 outlets or 6 duplex receptacles.

Take a long look at the minimum of two circuits for small appliances and consider adding more than the two circuits, even though two looks like enough. In this line of thinking, refer to Figure 8-6; you will observe that wiring circuits Nos. 7 and 13 have been run into the kitchen, with No. 13 having one duplex receptacle in the dining room, but there is also circuit No. 8 in the dining room.

Looking at the basement plan in Figure 8-7, you will notice that it does not indicate what the room with four lights is to be used for. It might be a bedroom, and if so, the receptacle outlets are adequate; but if it is a family room or recreation room, then the receptacle in the closet must be put on a lighting circuit, because it would not be a small appliance receptacle.

In Figure 8-6, you will notice two weatherproof receptacles (WP), one by the front door and one by the rear door of the home. These outlets are required to have *ground-fault circuit interrupters* (GCFIs) on the circuits because they are outdoor receptacles. They could both be put on the same circuit. GFCIs are also required for bathroom receptacles, in garages, and for countertop receptacles within six feet of the kitchen sink (see *Section 210.8*).

An arc-fault circuit interrupter (AFCI) is a device that is similar to a GFCI. An AFCI protects persons and equipment from an arc fault

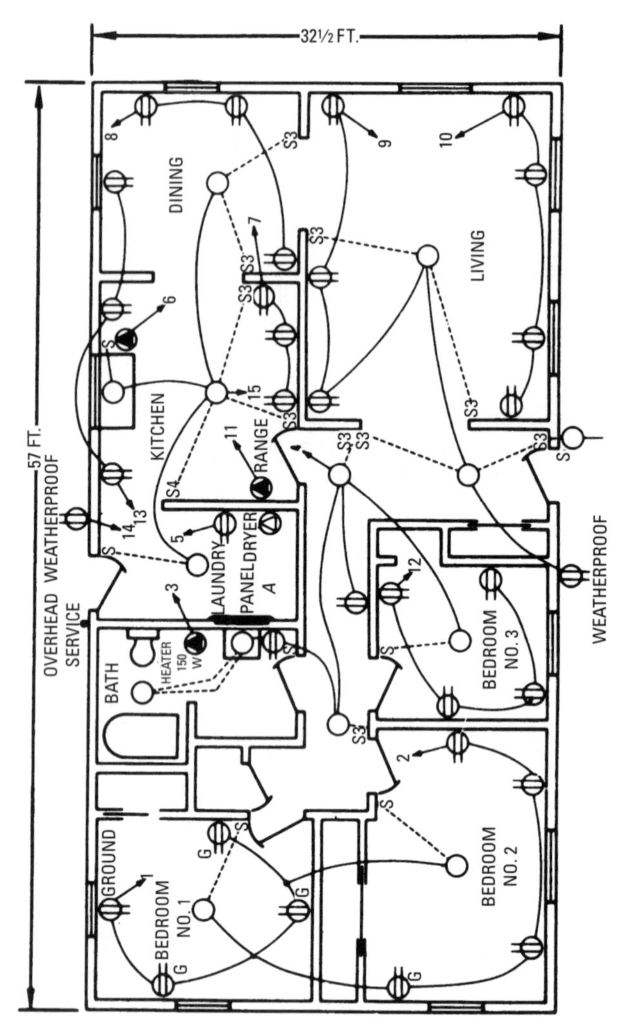

Figure 8-6 Circuiting of a house.

by recognizing the characteristics unique to an arcing fault and deenergizing the circuit when an arc fault is detected. All branch circuits supplying 15- or 20-ampere, single-phase 125-volt outlets installed in dwelling unit bedrooms must be AFCI protected by a listed device that protects the entire branch circuit. Recall that the definition

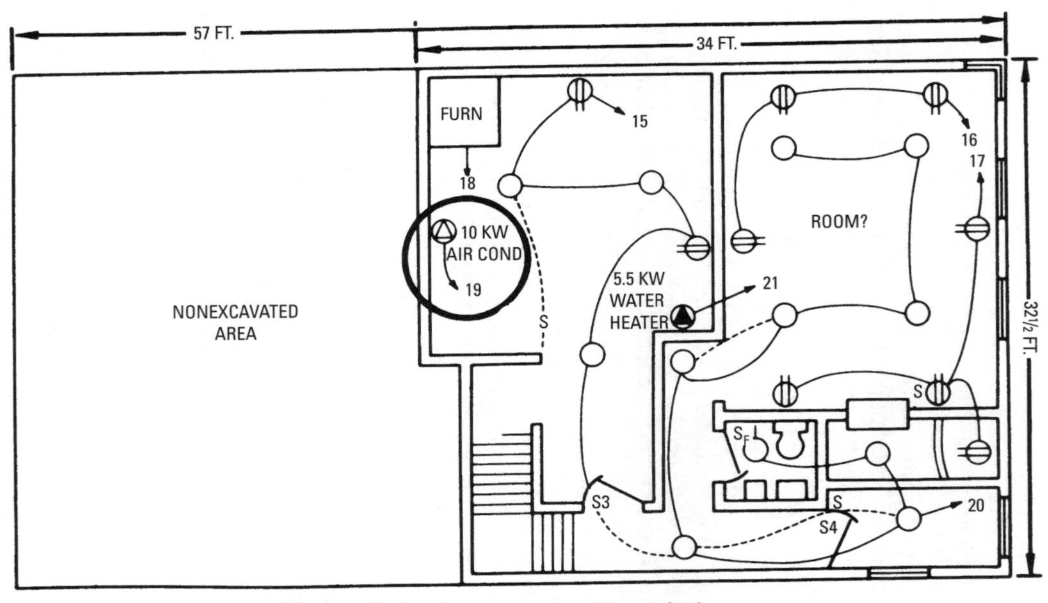

Figure 8-7 Basement circuiting of a house.

of an outlet at the beginning of this section includes openings for luminaires as well as for receptacles.

The traditional practice of separating the lighting from the receptacle circuits in dwelling unit bedrooms will, since the addition of this requirement, require two AFCI circuit breakers. The 125-volt limitation to the requirement means AFCI protection is not required for a 240-volt circuit, such as one for an electric heater. AFCI is also not required where it might diminish the overall safety of the installation. For instance, AFCI protection can be omitted from a smoke detector even if it is in a prescribed area for protection.

In Figure 8-7, we see a central air conditioner on circuit No. 19. This may not be installed—the customer may wish to have wall or window air conditioning. If this should be the case, circuit No. 19 could be eliminated and a 120- or 240-volt receptacle circuit run to the location where the wall or window air conditioner will be installed. If the circuit or circuits are for air conditioning only, the circuit may be loaded to 80 percent of its ampacity, but if other loads, such as lighting, are on these circuits, the air conditioner may load the circuit only to 50 percent.

In this summary, nonmetallic-sheathed cables will be the principal wiring method shown, because the majority of homes are wired with NM cable. In most instances, wiring with rigid metal conduit or electrical metallic tubing will be very similar except for the mechanics of installation and the fact that conductors are pulled in after the raceway is installed.

Circuits

Three-way and four-way switches seem to cause a little trouble in their connections. Three-way switches are used for turning lights on or off from two places. Three-way and four-way switches are used for turning lights off or on from three or more points (Figures 8-8 and 8-9).

Referring to Figure 8-9, if more than three switching points are required, merely add the extra switches (four-way) in the circuit between the two three-way switches. Often the switching of lighting at the garage is to be done from either the house or the garage. Also, sometimes a hot receptacle may be required in the garage. This connection requires a little more effort, so Figure 8-10 is included as a schematic for such a connection.

The fifth conductor (equipment-grounding conductor) may be eliminated by installing a fuse or breaker box in the garage for the receptacle(s) circuit. A made electrode is driven in the ground at the garage for the common grounding conductor and connected

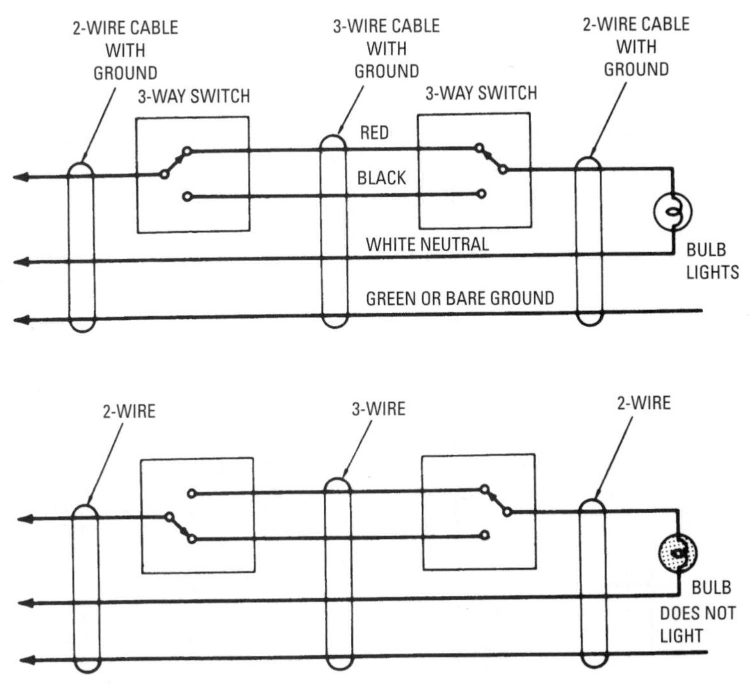

Figure 8-8 Three-way switch connections.

to both the neutral bus of the overcurrent device in the garage and the breaker enclosure (Figure 8-11).

Conduit

Most homes are wired with NM cable, but even if the house is wired with NM cable a certain amount of conduit or EMT usually must be installed. The wiring of basement walls for outlets should be laid out and conduit installed while the concrete forms are being set for the pouring of the walls. In setting the outlet boxes, do not be concerned with any furring strips used on the walls when they are finished. Box extensions can be added to take care of the furring. When installing conduits for outlets in basement walls, the same outlet spacing will be required. The basement may not be finished at the time of occupancy, but it will be later—of this you may be assured.

Figure 8-12 illustrates how to rough-in conduit before pouring the concrete. The conduit needs only to be stubbed out above the plate as it will be a raceway for NM cable. Stuff the box with

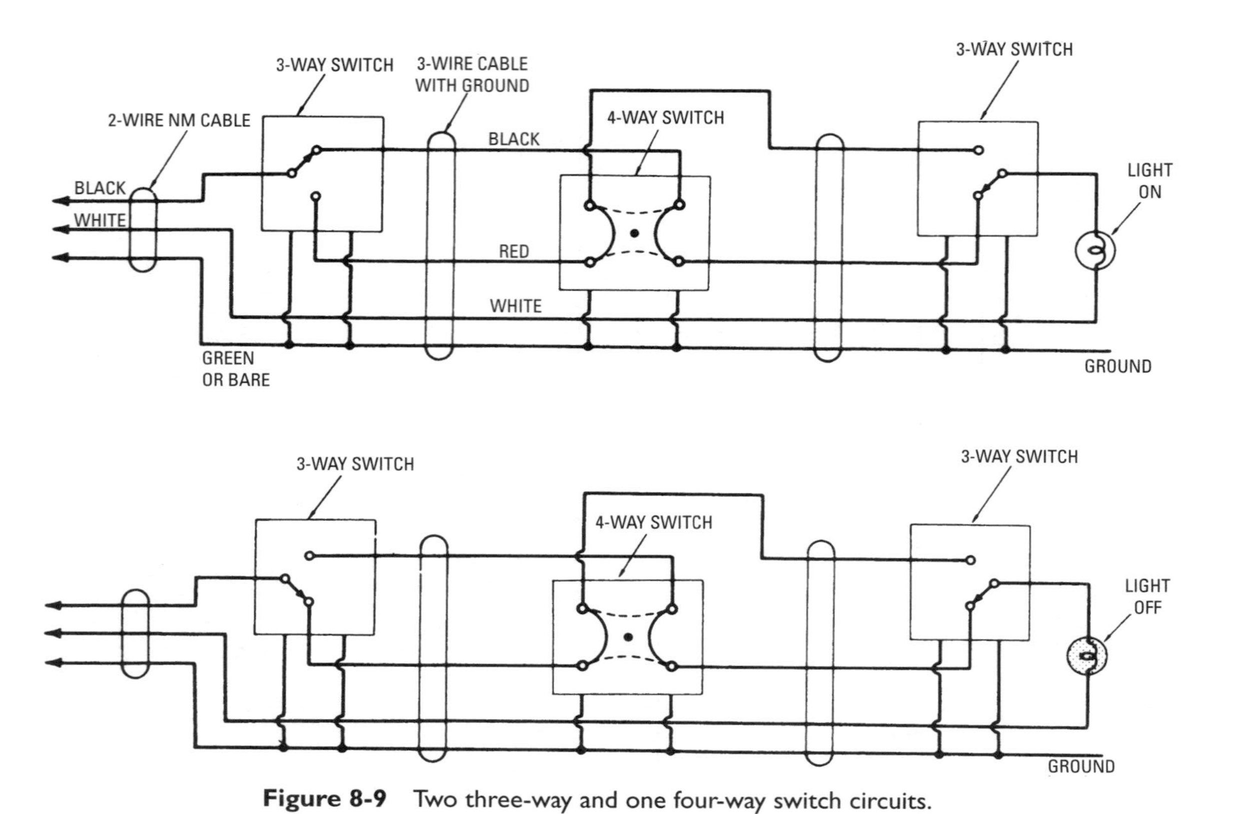

Figure 8-9 Two three-way and one four-way switch circuits.

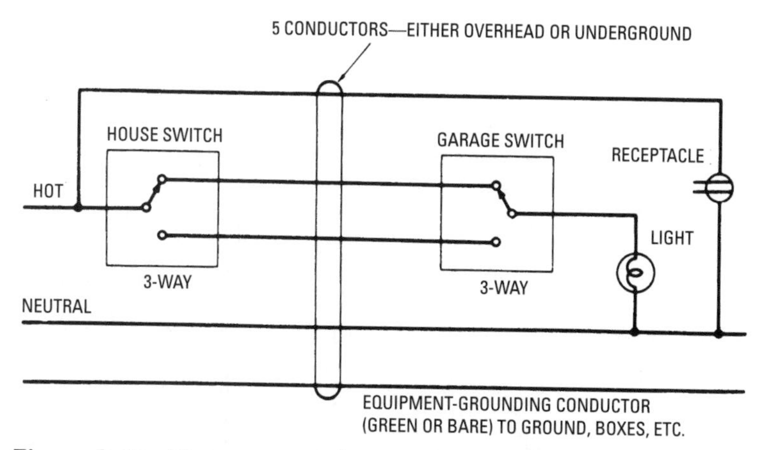

Figure 8-10 Three-way switch circuit for garage.

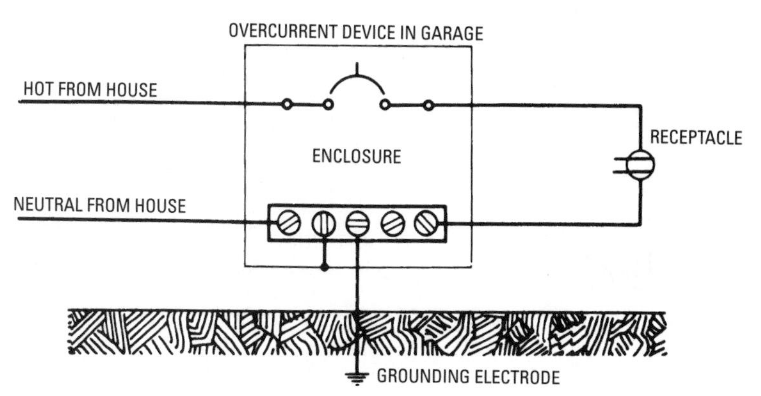

Figure 8-11 Installing a breaker box in garage.

newspaper to assist in cleaning out any concrete that may seep into the box. Conduit may also be run horizontally between outlets, if you so desire.

At most outlets in concealed work, where a house is wired with conduit, a right-angle bend is necessary. The inexperienced person has difficulty in making this bend to proper length. In general, the conduit to be bent must have a total depth from the back of the pipe to the end of the bend. The procedure is as follows: Secure a piece of conduit and, by the aid of a hickey, bend the end up slightly from the floor, keeping your left foot on the conduit and close to the hickey.

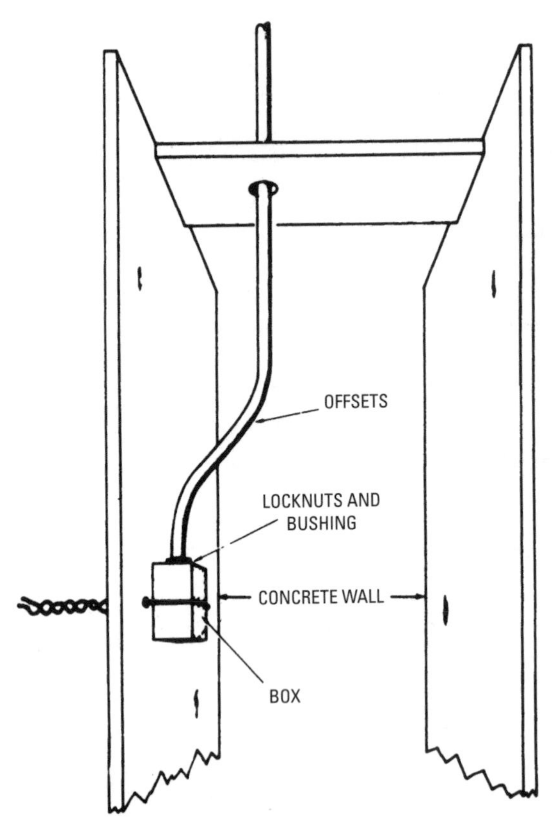

Figure 8-12 Conduit and box installation installed before pouring concrete wall.

Exert your bending force in two directions: one toward the bend along the line of the hickey handle and the other toward the right foot. When the bend is about three-fourths completed, measure up from the floor to the end of the conduit to determine whether the bend is going to be too short or too long when completed. If it is necessary to shorten or lengthen the bend, you can do so at this time by sliding your hickey up or down on the conduit and continuing to bend, being sure to apply the various forces as directed.

Great care should be taken in order to get all conduits and boxes lined up so that, when the mechanical work on the home is completed, the electrical devices installed will present a neat appearance. Before conductors are installed in the conduit, it should be swabbed out if it is found to contain water. Remember, no conductors are to be installed in conduit until all of it is complete and in place and the home is protected from the elements by the roof and walls.

As a rule, conduit is not used to any great extent in a frame home. When conduit is installed and concealed, notches are cut in the upper side of the floor just large enough to receive the conduit. *Do not* weaken the floor joists. Care must be taken not to recess the joists and other framing members beyond the amount really necessary, because of the weakening effect.

A conduit installation is more expensive than an NM cable installation. However, it has the advantage of being more flexible, because conductors may be removed and new conductors substituted or added at any time, provided that the conduit fill has not already been exceeded and that any derating required by *Table 310.15(B)(2)(a)* of the *NEC*, reproduced as *Table 4-5* in this book (and applied to *Tables 310.16* through *310.19* of the *NEC*), is observed.

Refer to Chapter 4 for coverage of electrical metallic tubing (EMT) and rigid metal conduit, and abide by what is covered in these two chapters. Pay particular attention to Figure 4-31, on page 90, regarding kinks and wrinkles. Figures 4-28 and 4-29, on page 89, pertain to the number of bends between pull boxes. Unless you have experienced trouble with pulling conductors or stripping insulation when pulling conductors, this rule may not mean too much to you. Yet, if the rules such as the limit on the number and total amount of bends are not adhered to, you will regret your new experience, and it will be rather late to rectify your mistakes; besides, you always have an inspector watching for *Code* violations. Always do a good job of reaming conduit and EMT ends, as shown in Figure 4-32, on page 91.

Circuit Layout

You have, or should have, a print of the house and the layouts of circuits and outlets on the print. You should also ascertain at what height your customer wishes the outlets and switches to be installed. Check to be sure that the layout of outlets and lighting meets the approval of the home owner. Stay within spacing required by the *NEC*, but add more outlets if the customer so desires.

In discussing the heights of receptacles and switches with your customer, the following heights seem to be the most used:

Receptacles: 12 inches to 18 inches

Receptacles over countertops: 42 inches

Receptacles in bathrooms over or by sink: 42 inches

Switches: 42 inches to 54 inches

After you have the height information for outlets and switches, use a 5-foot-long 1 × 2 board and mark it off for use in measuring the height of outlets and switches (Figure 8-13). This template, or marking stick, is then used to mark the location of the outlet box. After the outlets have been spotted for location, use this board and mark the height on the studs. Nothing looks worse than *X* number of receptacles in a room at *X* number of heights.

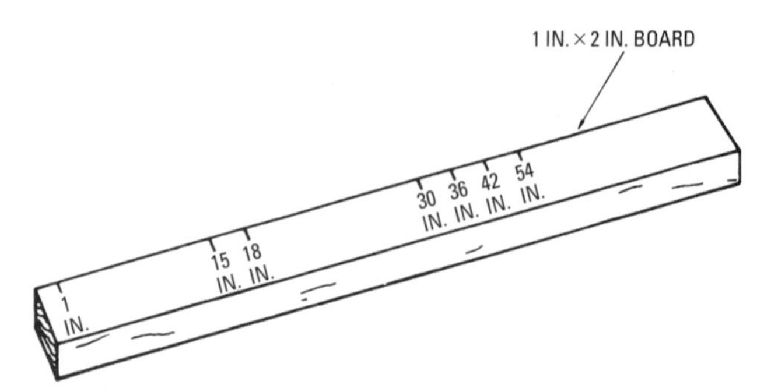

Figure 8-13 A template used to mark outlet and switch heights.

Now that you have locations and heights established, what direction will you run the cable to feed these outlets? It will be impossible to give you a hard-and-fast rule for this, but suggestions should be made. Long runs of cable will certainly work, but the longer the run, the greater the voltage drop. The electrical energy lost, plus the extra footage of cable, will cost money—not only first cost but every day. Analyze the wiring layout; a little time spent in doing this will pay off.

The wiring in a house is like handwriting to an inspector. In most cases an inspector can tell you who wired the house from examining the wiring. Pride should be taken in your workmanship, and *Code* requirements must be heeded.

Junction boxes are necessary items. Some must, of necessity, be used in attics and crawlspaces, but any extra connections take time and are potential sources of trouble, so the fewer the connections, the better.

Go back and take a look at Figure 8-6. The receptacles in bedroom No. 1 were put on a circuit with the light in bedroom No. 2 and, by the same token, the receptacles in bedroom No. 2 were put on the same circuit with the light in bedroom No. 1. This would leave either the light or the receptacles in either bedroom operable in case of trouble on either circuit No. 1 or No. 2.

Notice that the receptacles are placed so that no place along the wall is more than six feet from a receptacle. This is a *Code* requirement.

You will also notice that the furnace is on circuit No. 18, in Figure 8-7. It is not a *Code* requirement that the furnace be on a separate circuit, but if the furnace circuit had other outlets on it and trouble developed on those outlets, the furnace would not operate and the home would be without heat. A fuse at the furnace, sized to the value required by the furnace motor, should be installed. As to the dishwasher and disposal, they shall not be put on small appliance circuits. The author would suggest a multiwire circuit as shown in Figure 8-14.

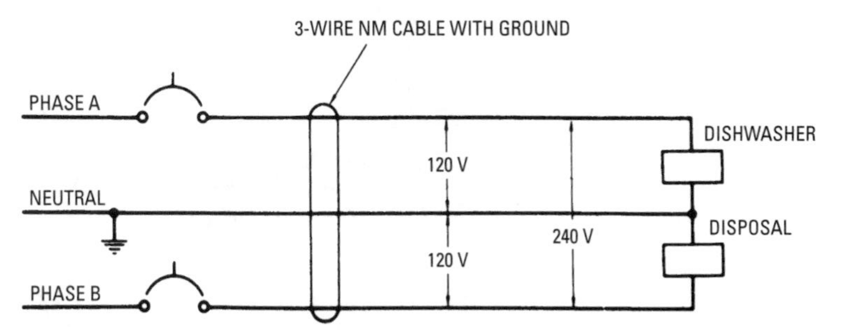

Figure 8-14 Multiwire circuit for dishwasher and disposal.

This is the *NEC* definition of a multiwire branch circuit: *a branch circuit consisting of two or more ungrounded conductors having a potential difference between them, and a grounded conductor having equal potential difference between it and each ungrounded conductor of the circuit and which is connected to the neutral (grounded) conductor of the system.*

The dishwasher is on one 120-volt circuit, direct-connected and well grounded. Exceptions for this are portable dishwashers, which will be plugged into small appliance circuits. The disposal is on the other 120-volt circuit and is also well grounded.

One item to remember is that a disposal is often removed for repairs by someone other than an electrician. If it is directly connected to the circuit, someone not familiar with equipment-grounding conductors could very easily not appreciate the value of the equipment-grounding conductor and miss the proper connection. Personally, the author prefers to see the disposal connected by cord and grounding-type attachment plug for easy removal and replacement, as shown in Figure 8-15. This method, of course, may not appeal to you. If not, direct-connect it.

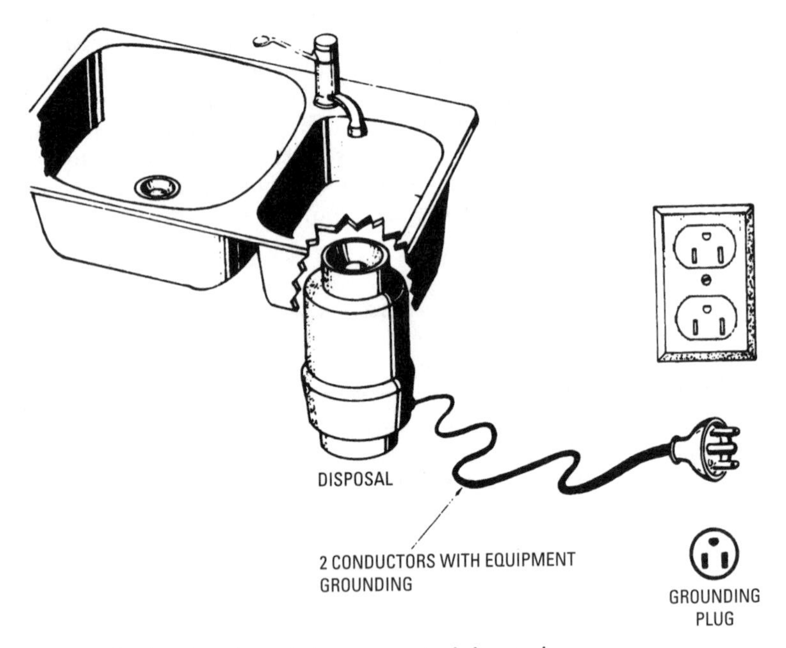

Figure 8-15 Cord- and plug-connected disposal.

Color-Coding

Section 200.7 of the *NEC* requires that white- or gray-colored conductors are to be used for identified (neutral) conductors and shall be used for no other purpose. *Section 200.6* of the *NEC* specifies that insulated conductors larger than No. 6 shall have an outer identification of white or gray color or shall be identified by distinctive white marking at terminals during the process of installation. The reason for permitting marking of conductors larger than No. 6 is that white or gray insulation is generally not available on No. 4 and larger conductors (Figure 8-16).

There are exceptions for NM sheathed cable and AC cable, because two-wire cable has one black and one white conductor and three-wire cable has one black, one red, and one white conductor. See cable markings in Figures 8-17 through 8-21.

Figure 8-16 Identification of conductor cable No. 4 and larger.

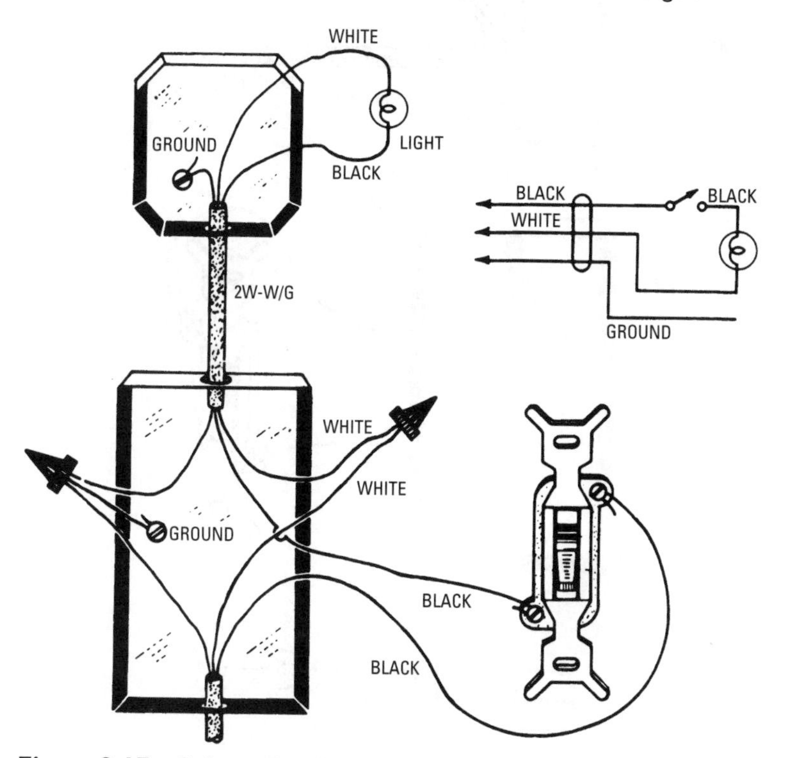

Figure 8-17 Color-coding for circuit with single-pole switch and light.

From these illustrations, you can readily see that white (neutral) goes to the light or the screw connection on screw-base sockets. Elsewhere, the white may be used as a traveler as in Figure 8-18, but then a black wire shall go back to the light. A neutral (identified conductor) is never connected to a switch. There is nothing to prohibit you from color-coding off colors at switching points; in fact, it is recommended.

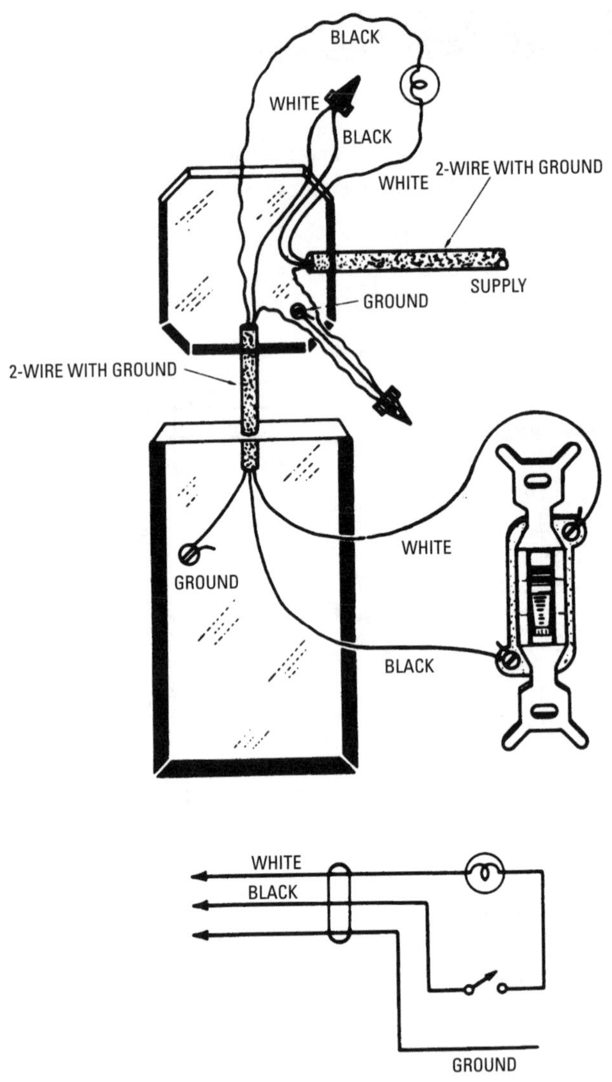

Figure 8-18 Color-coding for circuit with light at feed and a single-pole switch.

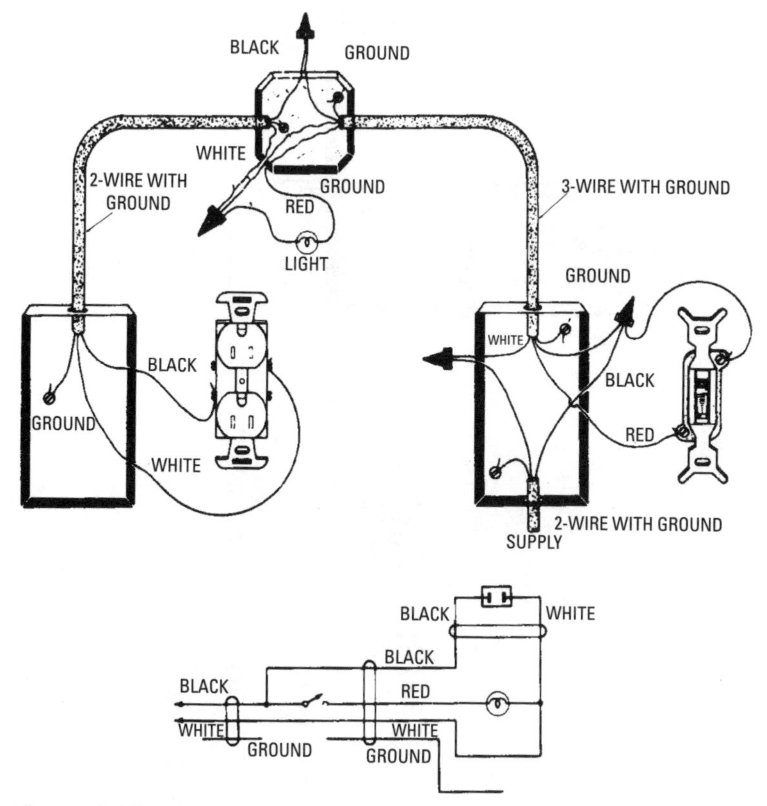

Figure 8-19 Coding for a switched light and hot receptacle.

Box Fill

It might be well to review box fill as described in Chapter 4, using Figures 8-17 through 8-21. In Figure 8-17, we have two 2-wire cables with equipment-grounding conductors connected to the switch box. Thus, we have total of four current-carrying conductors entering the box, plus two equipment-grounding conductors, making a total of six conductors. Then we have cable clamps and one device (switch). The two grounding conductors are counted as one conductor, plus four current-carrying conductors equals five conductors. We add the cable clamps and two conductors for the switch, making eight conductors total for the purposes of figuring fill. If these are No. 12 conductors, using *Table 4-6*, we find that it will take a $4 \times 1^1/_4$ square box.

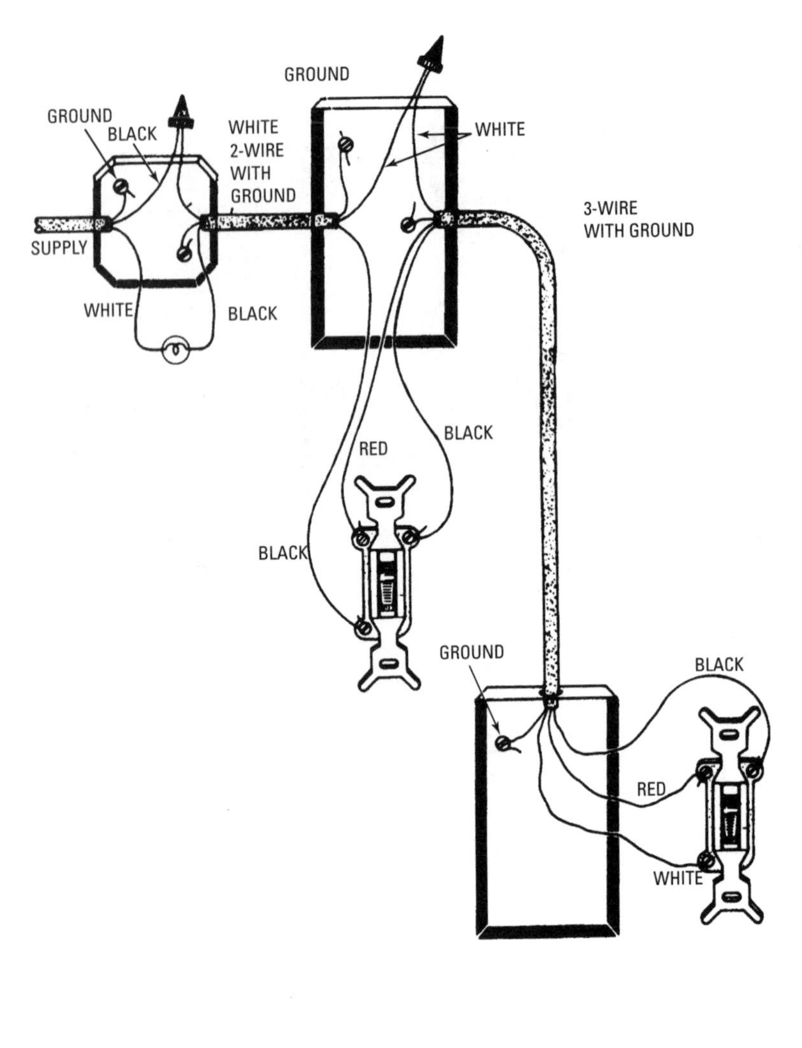

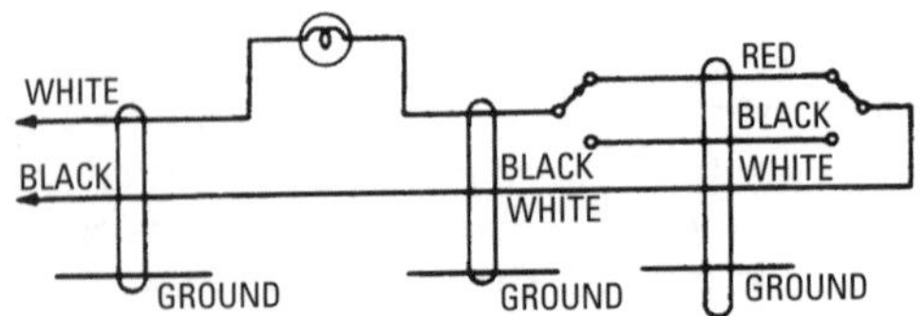

Figure 8-20 Coding for feed to light with two 3-way switches.

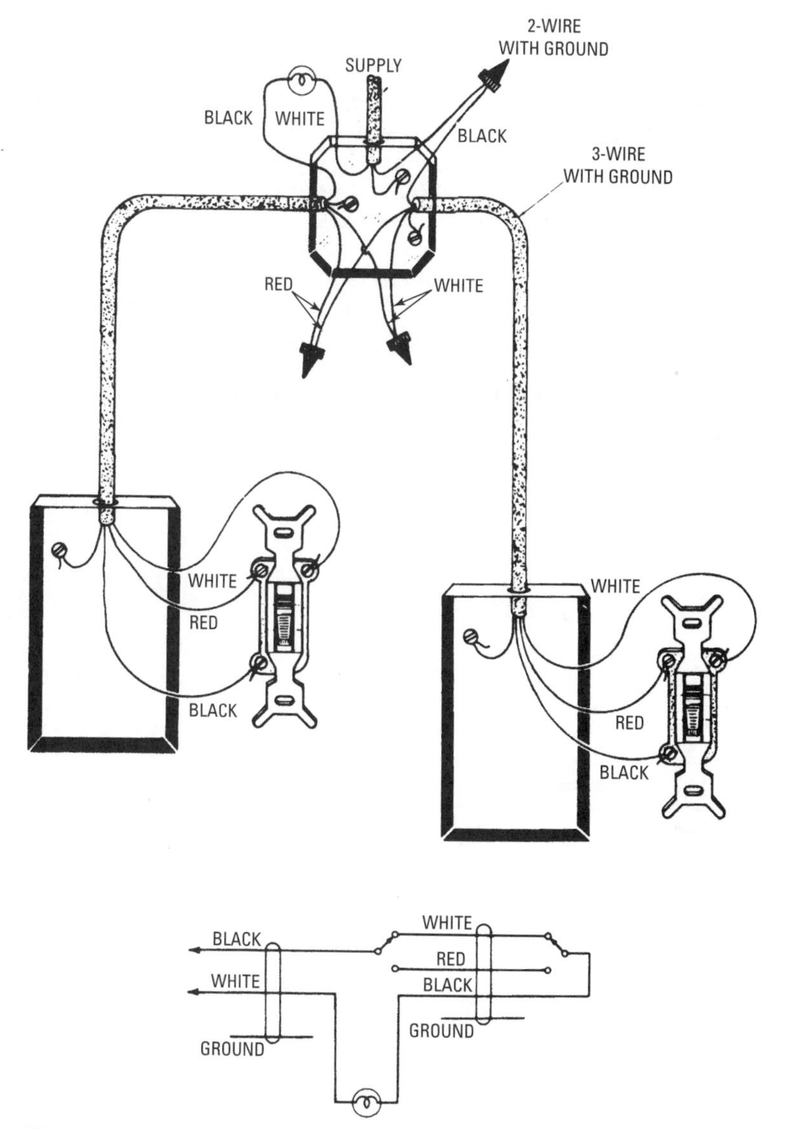

Figure 8-21 Coding for feed to light with two 3-way switches.

In the octagon box for the light, we have two-wire cable with ground, or three conductors, and cable clamps. There would be no device, but if a fixture stud were used for mounting the light fixture, we would have three conductors, cable clamps, and a fixture stud, making a total of five conductors to use in calculating the box size. With No. 14s, using *Table 4-6,* we could use either a 4 × 1¼ octagonal box or a 3 × 2 × 2 device box. If No. 12s were used, we could use the 4 × 1¼ octagonal box or a 3 × 2 × 2½ device box.

This does not mean that you must use the exact box; you may use larger boxes, but never smaller boxes. Because boxes for wiring a home are purchased in quantity, you may prefer the larger box to simplify the number of different size boxes that you use.

Consider the octagon box in Figure 8-21. Here we have two 3-wire cables and one 2-wire cable, with grounds. Thus, we have eight current-carrying conductors and three grounding conductors, but we are only required to count one grounding conductor, so we have nine conductors to figure—plus cable clamps and fixture stud, if used, or eleven conductors.

If No. 14s are used, from *Table 4-6,* we find we need a 4¹¹⁄₁₆ × 1¼ square box. If No. 12s were used, we would need the same size square box.

Installation of Cables

Allowance must be made to allow cable movement as a house settles. In Figures 8-22, 8-23, and 8-24 you will notice that the staples are

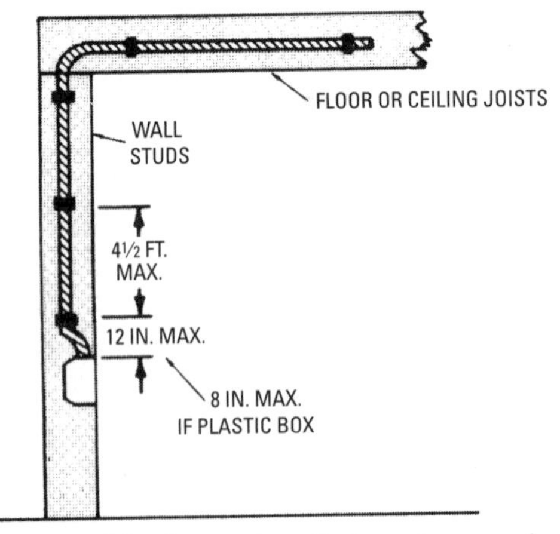

Figure 8-22 Proper installation of exposed cable.

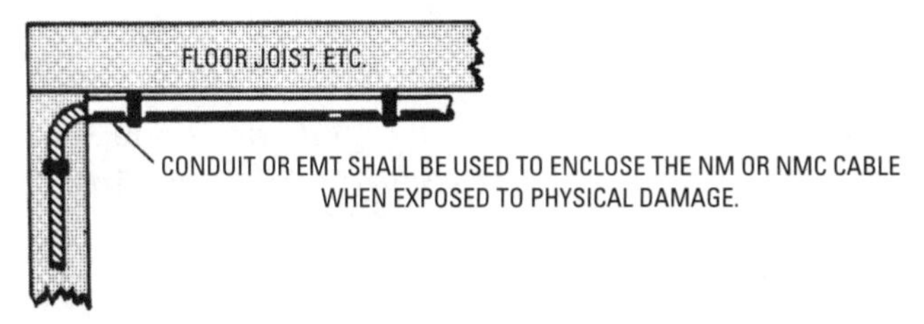

Figure 8-23 Installation to protect cables from physical damage.

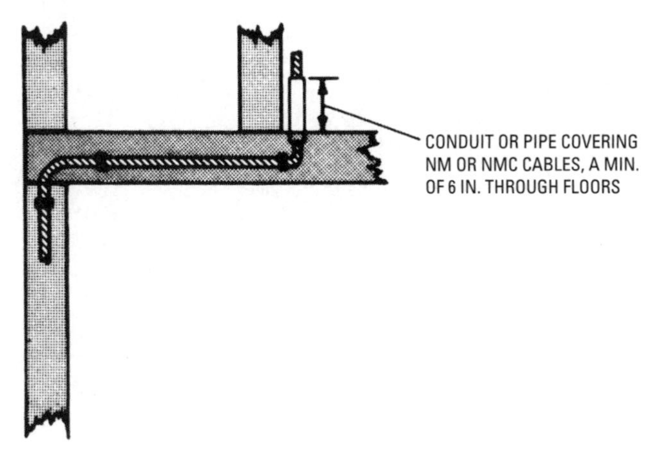

Figure 8-24 Protecting cable where it passes through a floor.

not placed right at the cable bends. This precaution allows for freedom of movement. Also, never install the staple too tight, because you might cut the insulation.

Most homes have an attic scuttle hole for access to the attic. *Section 320.23*, covering AC cable, will be quoted here because it also applies to NM sheathed cable:

> *320.23. In Accessible Attics. Type AC cables in accessible attics or roof spaces shall be installed in (A) and (B) below.*
>
> *(A) Where Run Across the Top of Floor Joists. Where run across the top of floor joists, or within 7 feet (2.13 m) of floor or floor joists across the face of rafters or studding, in attics and roof spaces which are accessible, the cable shall be protected by substantial guard strips which are at least as high as the cable. Where this space is not accessible by permanent*

stairs or ladders, protection will only be required within 6 feet (1.83 m) of the nearest edge of the scuttle hole or attic entrance.

(B) Where Carried Along the Sides of Floor Joists. Where cable is carried along the sides of rafters, studs, or floor joists, neither guard strips nor running boards shall be required.

From Figures 8-25 and 8-26, you see that there are two alternatives: to keep the cable a minimum of 6 feet from the scuttle hole or, if the run is closer than 6 feet, to install running boards for protection of the cable.

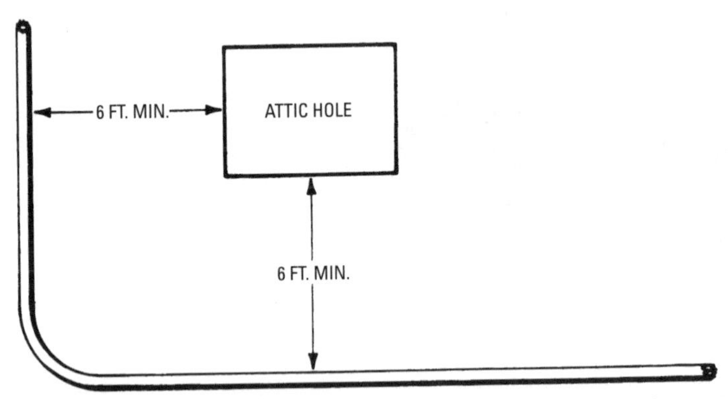

Figure 8-25 Cable in attic not protected by running boards.

Box and Device Grounding

We should never lose sight of the importance of the equipment-grounding conductors. Therefore, we must take every precaution to see that the equipment-grounding conductors are electrically continuous and properly made, with low impedance, so that they may serve the purpose for which they were intended.

What happens if a phase conductor shorts to ground in a box, or an electric drill, hedge clipper, saw, or other appliance develops a ground? The box or appliance housing develops a potential above ground that will be discharged through the body of whoever is holding the appliance or touching the box. The equipment-grounding conductors are paths for this potential (above ground) to return to the fuse or breaker protecting the circuit involved, blowing the fuse or tripping the breaker, and thus removing this potential above ground.

In making up grounds in boxes where you have more than one grounding conductor, do not place more than one grounding

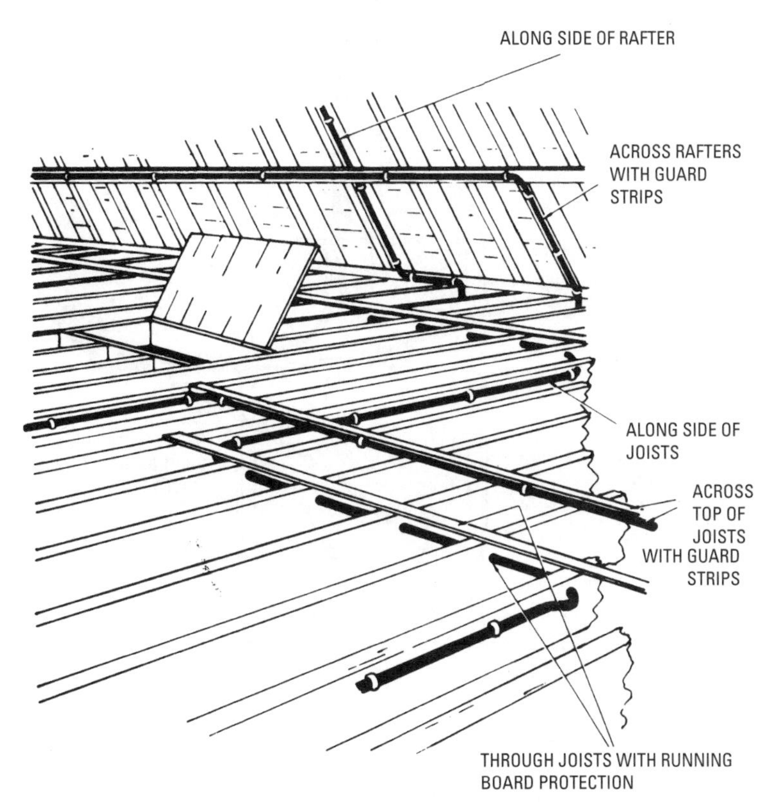

Figure 8-26 Method of running nonmetallic-sheathed cable in attic.

conductor under a box grounding screw. Splice all conductors together by means of an approved connector, installing a pigtail for the box grounding and another pigtail for the connection to the green grounding screw of the switch, receptacle, or other device (Figure 8-27). All boxes have tapped holes for $^{10}/_{32}$ grounding screws. Purchase the approved large-headed grounding screws—never use solder.

There are UL-approved devices such as receptacles that ground to the box and for which a grounding conductor will not be required. There are also UL-approved grounding clips used for grounding the equipment-grounding conductor to the box edge.

Switches and dimmers must also be grounded, as must metal faceplates that attach to them. If the device does not have a metal yoke that is in contact with a grounded metal box, a separate grounding conductor must be used.

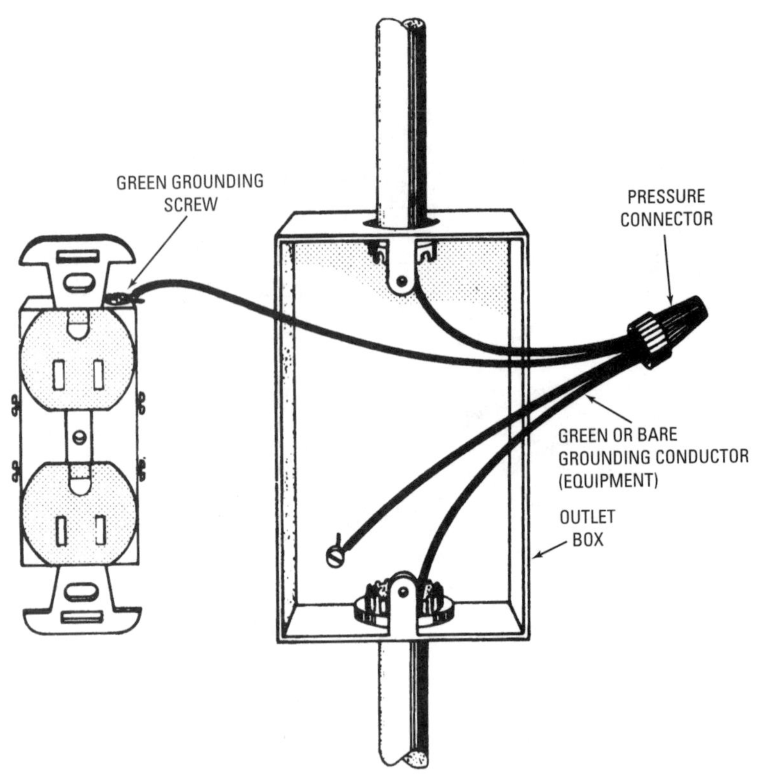

Figure 8-27 Proper connections for equipment-grounding conductors.

Wiring Basements

Earlier in this chapter, rough-in boxes in concrete basement walls were discussed. That discussion of course covered houses under construction. Sometimes you may be faced with installing outlets in a finished basement (concrete) wall.

If conduit was installed in the concrete walls during the construction, you will find the job of wiring simple. If there is no conduit, the wiring must be installed flush with the wall. Many people like to fur out the basement walls and install drywall or paneling. This will simplify the installation of the necessary outlets to meet the *NEC* requirements.

Furring is usually accomplished by nailing 1- or 1½-inch strips to the concrete or block walls, and the drywall or paneling is installed on these strips (Figure 8-28). Chisel out enough concrete to accommodate a device box of the proper cubic-inch capacity and

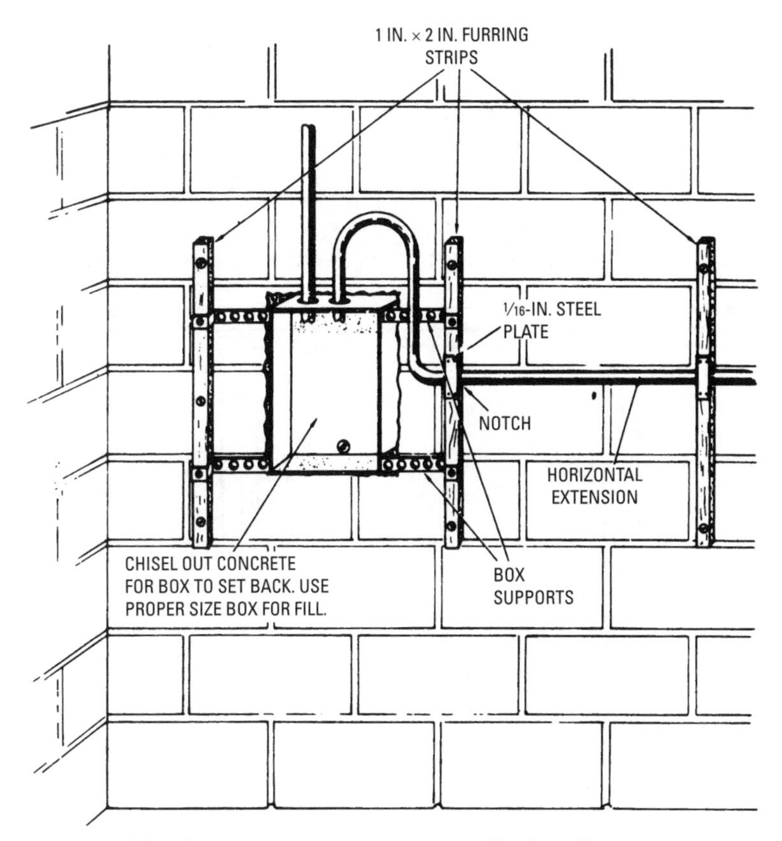

Figure 8-28 Furring strips on concrete or block walls.

support the box with box hangers, which will be required because of the damp location.

You may also continue with horizontal runs by cutting out notches in furring strips for the cable. Cover the cutout portion with 1/16-inch steel plate to protect the cable from nails.

Adding Wire to Old Homes

Here is where you use your ingenuity. No book could be large enough to cover all possible problems that you might encounter. Some of the more common problems will be covered.

If the older house has lath and plaster, the old plaster is often removed and replastered, or drywall is installed over the old plaster.

In these cases your problem will not be very large, as you may cut holes in the walls and fish the NM cable in very readily.

A very common problem that one runs into is headers (fire stops) between joists. The way this is usually overcome is illustrated in Figure 8-29. The header is located and plaster chipped out. The header is notched and covered with a $^{1}/_{16}$-inch steel plate after the NM cable has been installed and then patched with plaster. If the wall is brick, install NMC cable or UF cable at the bottom of the chase, install a strip of $^{1}/_{16}$-inch steel over the cable, and replaster.

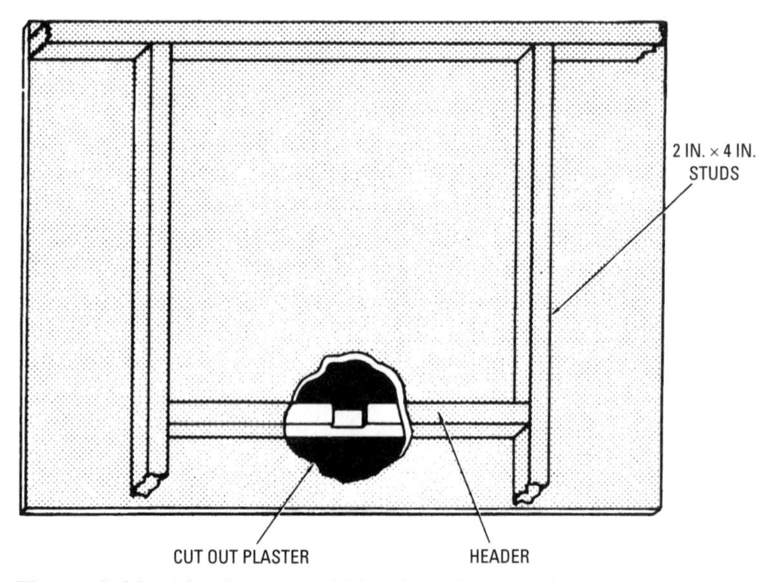

Figure 8-29 Notching out old headers (fire stops) for new electrical wiring.

Should you wish to fish down partitions between wall studs, it is generally best to use a small metal sash chain because of its flexibility. Locate the point at which you wish to install the outlet or switch, drill the top plates from the attic, and drop the chain through this hole. It usually can be heard rubbing the wall. Cut an opening, fish the chain with a bent piece of wire, and pull the nonmetallic-sheathed cable in.

Bear in mind that added wiring and devices must be grounded properly. Refer to Figure 4-4 for receptacle grounding. If new switches must be added or if a ground is not available, a nonconducting faceplate must be used. Or you may use a GFCI device to provide protection at the switch.

Appendix A

Calculations for Dwellings

The sample calculations in this appendix will be on the basis of *Table 220.11* of the *NEC* and Chapter 3 of this book. This method is used for determining the proper size of electrical service to a house. The goal of these calculations is that the service to the house be capable of supplying enough current to operate all of the loads (power-using devices) in the residence. The figures, such as watts-per-square-foot, have been developed over many years in the industry. The heading of *Table 220.11* is *Lighting Load Demand Factors*. Do not let this confuse you; it is actually for feeder and service calculations.

Calculation No. 1

Let us use the dimensions of the residence illustrated in Figure 3-3, parts A and B, of Chapter 3. This is necessary because the load calculations are based upon the square footage of the house. The outside dimensions of the ground floor are 32½ feet by 57 feet, giving an area of 1,852½ square feet. The basement measured 32½ feet by 34 feet, or 1,105 square feet in area. Dimensions of less than 6 inches and small offsets can be ignored, because they do not materially influence the final calculations. Assume that the basement will be finished; if not finished at the time of original construction, basements are often finished at a later date. Thus there is a total area of 2,957½ square feet.

There will be a 12-kW range and a 6.9-kW electric dryer. For those who are not familiar with kW, 1 kilowatt (kW) is 1,000 watts, and 1 watt is 1 volt times 1 ampere.

The general lighting calculation based on these figures is shown in *Table A-1*.

The 21,530 watts has to be now broken down into amperes per phase. A single-phase, 3-wire, 120/240-volt service will be used. Watts divided by volts equals amperes, and since this is single-phase 3-wire, the loads should be divided so that the amperage in each phase conductor is balanced insofar as is possible. Figure A-1 illustrates the voltages present in a service of this type. The two phases are designated by the letters *A* and *B*, and the neutral or grounded conductor by the letter *N*. Dividing 21,530 watts by 240 volts equals 90 amperes.

> *230.23. Size and Rating.*
>
> *(A) General. Conductors shall be of sufficient size to carry the loads as computed in accordance with Article 220, and shall*

Table A-1 General Lighting

2,957½ sq. ft. @ 3 watts per sq. ft.	8,872½ watts
Minimum of 2 small appliance circuits @ 1,500 watts per circuit	3,000 watts
Minimum of 1 laundry circuit @ 1,500 watts	1,500 watts
Total general lighting load	13,372½ watts
From *Table 220.11*:	
3,000 watts @ 100%	3,000 watts
13,372½ watts – 3,000 watts = 10,372½ watts @ 35%	3,630 watts
	6,630 watts
12 kW range (see *Table 3-1*)	8,000 watts
6.9 kW dryer (no demand factor; 100%)	6,900 watts
	21,530 watts

have adequate mechanical strength. Ampacity is determined from *Section 310.15.* This section will refer you to *Tables 310.16* through *310.19* and all applicable notes to these tables.

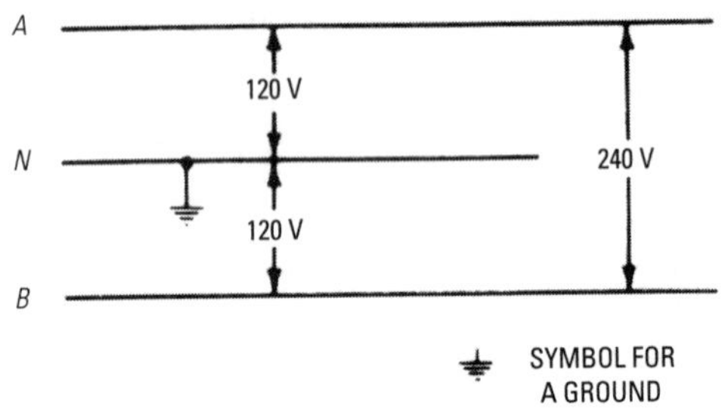

Figure A-1 A single-phase, 3-wire, 120/240-volt service.

(B) Minimum Size. The conductors shall not be smaller than 8 AWG copper or 6 AWG aluminum or copper-clad aluminum.

(C) Grounded Conductors. The grounded (neutral) conductor shall not be less than the minimum size as required by Section 250.24(B).

In Calculation No. 1, it was determined there were 90 amperes per phase. This is within the 100 amperes mentioned in *Section 310.15(B)(6)*. It would be wise to make the service entrance larger than 100 amperes so that the service size will not have to be enlarged in the near future.

Table A-2 Additional Loads for Garage, Outdoor Outlets, and Furnace Fan

Phase A	*Phase B*
90.0 amperes	90.0 amperes
5.2 amperes for fan	4.5 amperes for 3 outlets in garage
	3.0 amperes for 2 outlets on porch
95.2 amperes	97.5 amperes

Some addition to the original calculation, though not specifically required, is also suggested. There was no garage figured, nor any lights or receptacles for porches or a patio. General-purpose outlets are calculated at 1½ amperes per outlet, and the fan motor on the furnace (if one is present) should be included. These additions are listed in *Table A-2*.

The calculations just made will now be broken down into number and sizes of branch circuits.

General lighting was 2,957½ square feet at 3 watts per square foot, or 8,872½ watts. These will be 120-volt circuits, so 8,872½ divided by 120 volts equals 73.9 amperes. This would consist of a minimum of six 15-ampere circuits or four 20-ampere circuits for general lighting including general-purpose receptacles. These are the minimum number required, but consider adding a few more. There are also a minimum of two 20-ampere small appliance circuits (as explained in Chapter 3) and one 20-ampere laundry circuit.

Using the demand factor in *Table 220.19* for the 12-kW range, 8 kW was indicated for one appliance. So, 8,000 watts divided by 240 volts equals 33.3 amperes, calling for a 35-ampere circuit. Referring to *Table 310.16* and *310.18* of the *NEC* shows that this will require a minimum of No. 8 copper, No. 6 aluminum in 60°C (140°F) wire or No. 8 aluminum in 75°C (167°F) wire. It is recommended that wire no smaller than No. 6 copper, or the equivalent in aluminum, be used.

The dryer was rated at 6,900 watts, so 6,900 divided by 240 volts equals 29 amperes. No circuit should be loaded 100 percent. Since 80-percent loading is a safe figure, and since the loads of electric dryers have been increasing, use No. 8 copper or the equivalent in aluminum, and a 40-ampere circuit. Notice that 240 volts was used on the range and dryer calculations because both are 120/240-volt appliances.

Optional Calculations

Section 220.30 describes an optional calculation that will generally be used for homes that have considerable electrical equipment in them. This section of the *Code* is reproduced here and will be used as the basis for the calculations in the subsequent section.

> *220.30. Optional Calculations—Dwelling Unit.*
>
> *(A) Feeder and Service Load. For a dwelling unit having the total connected load served by a single 3-wire, 120/240-volt or 208Y/120-volt set of service-entrance or feeder conductors with an ampacity of 100 or greater, it shall be permissible to compute the feeder and service loads in accordance with this section, instead of the method specified in Part II of this article. The calculated load shall be the result of adding the loads from 220.30(B) and (C). Feeder and service-entrance conductors whose demand load is determined by this optional calculation shall be permitted to have the neutral load determined by Section 220.22.* [*Table A-3* specifies the loads.]

Table A-3 Optional Calculation for Dwelling Unit Load in kVA

Largest of the following four selections.

1. 100 percent of the nameplate rating(s) of the air conditioning and cooling, including heat pump compressors.
2. 100 percent of the nameplate ratings of electric thermal storage and other heating systems where the usual load is expected to be continuous at the full nameplate value. Systems qualifying under this selection shall not be figured under any other section in this table.
3. 65 percent of the nameplate rating(s) of the central electric space heating including integral supplemental heating in heat pumps.
4. 65 percent of the nameplate rating(s) of electric space heating if less than four separately controlled units.
5. 40 percent of the nameplate rating(s) of electric space heating of four or more separately controlled units.

   Plus: 100 percent of the first 10 kVA of all other load and 40 percent of the remainder of all other load.

> *The loads identified* [in *Table A-3*] *as "other load" and as "remainder of other load" include the following:*
>
> 1. 1,500 volt-amperes for each 2-wire, 20-ampere small appliance branch circuit and each laundry branch circuit specified in Section 220.16.

2. 3 volt-amperes per square foot (0.093 square meter) for general lighting and general-use receptacles.
3. The nameplate rating of all appliances that are fastened in place, permanently connected, or located to be on a specific circuit; ranges; wall-mounted ovens; countermounted cooking units; clothes dryers; and water heaters.
4. The nameplate ampere of KVA rating of all motors and of all low-power-factor loads.
5. When applying Section 220.31 use the largest of the following: (1) air conditioning load; (2) thermal storage and other heating systems that have continuous loads; (3) the 65 percent diversified demand of the central electric space heating load; (4) the 65 percent diversified demand of the load of less than four separately controlled electric space heating units; (5) the 40 percent diversified demand of the load of four or more separately controlled electric space heating units.

Quoting further from the *NEC*:

> *220.31. Optional Calculation For Additional Loads in Existing Dwelling Unit.*
>
> *Where the (existing) dwelling unit presently being served by an existing 120/240 volt or 208Y/120, 3-wire service, it shall be permissible to compute load calculations* [on the basis of *Table A-4.*]
>
> *Load calculation shall include lighting at 3 volt-amperes per square foot (0.093 square meter); 1,500 volt-amperes for each 20-ampere appliance circuit; range or wall-mounted oven and counter-mounted cooking unit, and other appliances that are permanently connected or fastened in place, at nameplate rating.*
>
> *If air-conditioning equipment or electric space heating equipment is to be installed, the following formula* [shown in

Table A-4 Load Factors from the *NEC Section 220.31*

Load (in kVA)	*Percent of load*
First 8 kVA of load at	100
Remainder of load at	40

Table A-5 Load Factors for Air Conditioning and Space Heating

Air conditioning equipment*	100%
Central electric space heating*	100%
Less than four separately controlled space heating units*	100%
First 8 kVA of all other load	100%
Remainder of all other load	40%

*Use larger connected load of air conditioning and space heating, but not both.

Table A-5] shall be applied to determine if the existing service is of sufficient size.

Other loads shall include:

- *1500 volt-amperes for each 20-ampere appliance circuit.*
- *Lighting and portable appliances at 3 volt-amperes per square foot (0.093 square meter).*
- *Household range or wall-mounted oven and counter-mounted cooking unit.*

All other appliances fastened in place, including four or more separately controlled space heating units, at nameplate rating.

The calculations in the following sections will be based on *Sections 220.30* and *220.31*.

Calculation No. 2

Here, the first example will be calculated on the optional basis shown in the preceding section, and then the figures will be checked. This house has neither electric heating nor air cooling, so the calculations are as shown in *Table A-6*.

Dividing 18,909 watts by 240 volts equal 79 amperes. Using the first calculation, we obtained 90 amperes, and using the optional calculation, we obtain 79 amperes.

Calculation No. 3

This calculation will be for a 2,000-square foot dwelling exclusive of garage and porches but including the basement. There will be a 1.2-kW dishwasher, a 1-kW disposal, 10 kW of space heating installed in six rooms (not a central heating plant), and a 6-ampere, 230-volt air conditioner (1,380 watts divided by 1,000 equals

Table A-6 General Lighting, Optional Calculation

2,957$\frac{1}{2}$ sq ft @ 3 watts per sq ft	8,872$\frac{1}{2}$ watts
2 20-A small appliance circuits @ 1,500 watts each	3,000 watts
1 20-A laundry circuit @ 1,500 watts	1,500 watts
1 12-kW range (nameplate rating)	12,000 watts
1 6.9-kW dryer (nameplate rating)	6,900 watts
	32,372$\frac{1}{2}$ watts
1st 10 kW of all other load at 100%	10,000 watts
Remainder of all other load at 40% (22,272$\frac{1}{2}$ watts)	8,909 watts
	18,909 watts

Table A-7 Optional Calculation Including Space Heating, Cooling, and Kitchen Devices

2,000 sq ft @ 3 watts per sq ft	6,000 watts
220-ampere small appliance circuits @ 1,500 watts	3,000 watts
120-ampere laundry circuit @ 1,500 watts	1,500 watts
112-kW range @ 100%	12,000 watts
15-kW water heater @ 100%	5,000 watts
11.2-kW dishwasher @ 100%	1,200 watts
11-kW disposal @ 100%	1,000 watts
10 kW of space heating, divided in six rooms	10,000 watts
	39,700 watts
1st 10 kW @ 100%	10,000 watts
29,700 watts @ 40%	11,880 watts
	21,880 watts

1.38 kW). Note that the space-heating wattage is larger than the air-conditioning wattage, so the air-cooling unit will not appear in the calculations shown in *Table A-7*.

Then, 21,880 watts divided by 240 volts equals 91 amperes. So this installation will require a minimum service of 100 amperes.

Calculation No. 4

The same figures used in Calculation No. 3 will be used here except for 8 kW of central electrical heating, 5 kW of air cooling, two 4-kW wall-mounted ovens, and one 6.5-kW counter-mounted cooking top.

As in Calculation No. 3, this will be a 2,000-square foot dwelling, including basement but excluding garage and porches. There will be two 4-kW wall-mounted ovens, one 6.5-kW counter-mounted cooking top, 8 kW of central heating, 5 kW of air cooling, one 5-kW water heater, one 1.2-kVA dishwasher, and one 1-kVA disposal. The air cooling is larger in Calculation No. 3, and the heating is a central plant. Referring back to *Table A-5*, the air cooling can be ignored because it is smaller than the space-heating load, and 100 percent of the central heating load is used because central heating does not have the diversity of six rooms on separate thermostats. Also, there are now two ovens and one cooking top instead of a 12-kW range. The calculations are shown in *Table A-8*.

Table A-8 Optional Calculation with Central Heating

2,000 sq. ft. @ 3 watts per sq. ft.	6,000 watts
2 20-amp. small appliances circuits @ 1,500 watts	3,000 watts
1 20-amp. laundry circuit @ 1,500 watts	1,500 watts
Central space heating	8,000 watts
Air cooling	5,000 watts
2 4-kW wall-mounted ovens	8,000 watts
1 6.5-kW counter-mounted cooking top	6,500 watts
1 1.2-kVA dishwasher	1,200 volt-amperes
1 1-kVA disposal	1,000 volt-amperes
1 5-kW water heater	5,000 watts
	45,200 volt-amperes
The 8 kW of central heating is subtracted before figuring the percentage and then added back later	– 8,000 watts
	37,200 watts
First 10 kVA of the 37,200 volt-amperes @ 100%	10,000 volt-amperes
Remainder of 37,200 volt-amperes, or 27,200 @ 40%	10,880 volt-amperes
Central heating @ 100%	8,000 volt-amperes
	28,880 volt-amperes

Dividing 28,880 volt-amperes by 240 volts equals 120 amperes. This will require a service larger than 100 amperes. *Section 240.6* lists the next size breaker or fuse (standard size) as 125 amperes. It is suggested that at least a 125-ampere service and a 125-ampere breaker be used.

Calculation No. 5

The instructions in *Section 230.31* of the *NEC*, reproduced earlier in this chapter in the section *Optional Calculations*, will apply to an existing 120/240-volt or 208Y/120-volt, 3-wire service to which an additional load is to be attached.

In this particular calculation, the present loads will be ignored and the following figures will be used to bring the wiring up to the 2002 *NEC* standards.

There are 1,500 square feet of finished residential area, a 12-kW range, and a 3-kVA air cooler. Calculation of these loads is shown in *Table A-9*. It is now necessary to use 100 percent of the air-cooling load, which will leave 21,000 volt-amperes of other load, as shown in *Table A-10*.

Table A-9 General Lighting Load

1,500 sq ft @ 3 volt-amperes per sq. ft.	4,500 volt-amperes
2 20-ampere small appliance circuits @ 1,500 volt-amperes	3,000 volt-amperes
1 20-ampere laundry circuit @ 1,500 volt-amperes	1,500 volt-amperes
1 12-kW range	12,000 volt-amperes
1 3-kW air cooler	3,000 volt-amperes
	24,000 volt-amperes

Table A-10 Allocation of Load between Air Conditioning and Other Loads

Air cooling	3,000 volt-amperes
First 8 kVA of other load	8,000 volt-amperes
Balance 21,000 −8,000 volt-amperes = 13,000 volt-amperes @ 40%	5,200 volt-amperes
	16,200 volt-amperes

Dividing 16,200 volt-amperes by 240 volts equals 67.5 amperes. In this case, the 60-ampere service will have to be increased in ampacity, so refer back to *Table 220.11* or *Section 220.30* and recalculate the service size. In doing so, it will be necessary to increase the service to the 100-ampere minimum.

Motor Loads

See Chapter 3 in this book and *Section 220.14* of the *NEC*. Where motor loads, such as air-cooling motors, furnace motors, pump motors, and disposal motors, are used, 125 percent of the full-load current rating of the largest motor is taken, plus 100 percent of the full-load current rating of the balance of the smaller motors, all added to the load for figuring the feeders and services. On branch circuits, use 125 percent of the full-load current rating of the motor or motors involved to calculate the conductor size for the branch circuit. This assumes only one motor to a branch circuit. If there are two or more motors, take 125 percent of the largest full-load current rating of other motors on the same branch circuit.

If the motor nameplate is in amperes instead of horsepower, use the full-load nameplate rating, but if the rating is in horsepower, refer to *Table 430.148* of the *NEC* to arrive at the amperes. In doing this, make note of whether the motors are 115 or 230 volt. The 230-volt motors will be added to both phase legs (A and B). If a motor is a 115-volt motor, its load will be on one phase leg only, and it will be necessary to balance the load on both phase legs as nearly as possible. Motor loads other than air-cooling and furnace motors do not actually appear too frequently in residences.

It is recommended that reference be made to the examples in *Chapter 9* of the *NEC*, where additional calculations will be found to check further the method used in calculating.

Appendix B

Farm Buildings

Information on farm buildings and farm services appears in *Sections 220.40* and *220.41* of the *NEC*. Calculations for farm buildings take into consideration the demand factors for farm buildings and services. There are basically two types of service installations: the main service, which goes to the dwelling and from which the farm buildings are served, and the farm service pole, from which the separate buildings, including the dwelling, are served by service drops, service laterals, or both. The metering for this latter type is on the farm service pole. In wiring the dwelling, the loads in the other buildings must be considered in figuring the service equipment if the meter is at the dwelling. If a pole is used as the metering point, then the sizing of the conductors to the meter must be figured.

Probably the best way to present this is to quote the *NEC* and then follow with a calculation as an example.

> *220.41. Farm Loads—Total. Where supplied by a common service, the total load of the farm for service entrance conductors and service equipment shall be computed in accordance with the farm dwelling load and demand factors specified in Table 220.41* [reproduced here as *Table B-1*]. *Where there is equipment in two or more farm equipment buildings or for loads having the same function, such loads shall be computed in accordance with Table 220.40* [reproduced here as *Table B-2*] *and may be combined as a single load in Table* [*B-1*] *for computing the total load.*

Table B-1 Method for Computing Total Farm Load (in the *NEC, Table 220.41*)

Individual Loads Computed in Accordance with Table 220.40 in the NEC [Table B-2]	*Demand Factor (Percent)*
Largest load	100
Second largest load	75
Third largest load	65
Remaining loads	50

To this total load, add the load of the farm dwelling unit computed in accordance with Part II or III of this article. Where the dwelling has electric heat and the farm has electric grain drying systems, Part III of this article shall not be used to compute the dwelling load.
Note: Computation of dwelling loads is covered in Appendix A.

Table B-2 Method for Computing Farm Loads for Other Than Dwelling Unit (in the *NEC, Table 220.40*)

Ampere Load at 240 Volts Maximum	*Demand Factor (Percent)*
Loads expected to operate without diversity, but not less than 125% full-load current of the largest motor and not less than the first 60 amperes of load	100
Next 60 amperes of all other loads	50
Remainder of other load	25

Example of a Farm Calculation

This example, shown as *Table B-3*, excludes the farm residence. That is, it is for farm equipment only, not for the farm's residence(s).

Table B-3 Farm Equipment Load Calculations

Load No. 1 (feed grinder and auger)	
5-hp, single-phase, 240-volt motor	28 amperes
1-hp, single-phase, 240-volt motor	8 amperes
Load No. 1 total	36 amperes
Load No. 2 (milk barn)	
Lighting	2,000 volt-amperes
Water heater	2,500 volt-amperes
Total	4,500 volt-amperes
4,500 volt-amperes divided by 240 volts	19 amperes
2-hp milker	12 amperes
1-hp cooler	8 amperes
Air conditioner	15 amperes
Load No. 2 total	54 amperes
Load No. 3 (chicken house)	
Brooder	3,000 volt-amperes
Lighting	450 volt-amperes
Total	3,450 volt-amperes
Load No. 3 total (3,450 watts divided by 240 volts)	14 amperes

In these computations, the largest motor is 5-horse power rated at 28 amperes. This will have to be increased by 25 percent of full-load current, so it will be 35 amperes, making Load No.

1 35 amperes + 8 amperes = 43 amperes. The computed load is summarized in *Table B-4*.

Table B-4 Farm Equipment Load Summary

Load No. 1	43 amperes
Load No. 2	54 amperes
Load No. 3	14 amperes
Total	111 amperes
First 60 amperes at 100%	60 amperes
Next 60 (or less) amperes @ 50%	25.5 amperes
Remainder @ 25%	(0) amperes
Computed load using *Table 220.11*	85.5 amperes

Thus, the service to supply the farm buildings will not be the 113 amperes, but will be 85.5 amperes.

The next example is where there might be a farm service pole and service drops to all of the loads including the dwelling. The single-family dwelling has a floor area of 1550 square feet exclusive of an unfinished attic and porches. It has a 12-kW range.

To compute residence load, see *Sections 220.11* through *220.19*. Also refer back to Chapter 3, which covered the computations for dwelling occupancies. The individual loads are computed in *Table B-5*.

Table B-5 Load Calculations for Residence and Equipment

General lighting load	
1,500 sq. ft. @ 3 watts per sq. ft.	4,500 volt-amperes
Small appliance circuits	3,000 volt-amperes
Laundry circuit	1,500 volt-amperes
Total without range	9,000 volt-amperes
3,000 volt-amperes at 100%	3,000 volt-amperes
9,000 volt-amperes – 3,000 volt-amperes = 6,000 volt-amperes at 35%	2,100 volt-amperes
Net computed load without range	5,100 volt-amperes
Range load (see *Table 220.19*)	8,000 volt-amperes
Net computed load with range	13,100 volt-amperes

(continued)

Table B-5 (*continued*) Load Calculations for Residence and Equipment

Load No. 1 total for 120/240 volt, 3-wire system feeders	
13,100 divided by 240 volts	55.5 amperes
Load No. 2 (feed grinder and auger)	
5-hp single-phase, 240-volt motor	28 amperes
1-hp single-phase, 240-volt motor	8 amperes
Load No. 2 total	36 amperes
Load No. 3 (milk barn)	
Lighting	2,000 volt-amperes
Water heater	2,500 watts
Total	4,500 volt-amperes
4,500 volt-amperes divided by 240 volts	19 amperes
2-hp milker	12 amperes
1-hp cooler	8 amperes
Air conditioner	15 amperes
Load No. 3 total	54 amperes
Load No. 4 (chicken house)	
Brooder	3,000 volt-amperes
Lighting	450 volt-amperes
Total	3,450 volt-amperes
Load No. 4 total (3,450 volt-amperes divided by 240 volts)	14 amperes

Table B-6 Total Load

Largest demand (residence) 56 amperes @ 100%	55.5 amperes
Second largest demand (Load No. 3) 54 amperes @ 75%	40.5 amperes
Third largest demand (Load No. 2) — with 25% of the 28 amperes for the 5-hp motor added: 28 + 7 = 35, or a total of 43 amperes @ 65%	28 amperes
Balance of the demand (Load No. 4) 14 amperes @ 50%	7 amperes
	131 amperes

The total load is computed in *Table B-6* using the demand factors from *Table B-1*: This load includes the dwelling and is served from a farm service pole and service drops to all buildings.

The total connected load for the farm would be 170 amperes, but using the demand factors in *Sections 220.40* and *220.41*, a service to handle 135 amperes would be installed. Bear in mind that this would not allow much expansion for future demand.

Appendix C

Audio and Home Theater Installations

This appendix contains the necessary information for the construction of home theater and audio systems. These systems operate at low voltages, so the hazards associated with them are greatly reduced, compared to electrical wiring for lighting and power. The hazards inherent in the construction process, however, remain.

Sound Systems

Residential sound systems come in a variety of types, but their operating characteristics are nearly identical from one to another.

The four main parts of a sound system are as follows:

- **Signal generators** are the source of the audio signal. Common signal generators are CD players, tape decks, radio tuners, and microphones.
- **Signal processors** are the intermediate step between the signal generators and the sound generators. These devices normally include preamps, switches, noise reduction units, equalizers, filters, delay units, mixers, monitors, and amplifiers. These devices all boost, diminish, color, clip, or in some way modify a signal. When designing sound systems, a great deal of the effort should be spent on signal-processing equipment, because there are so many choices, and because different combinations of such items provide drastically different effects. Each combination of signal processors produces a final sound that is a little bit different from that produced by any other combination.
- **Sound generators** are devices that change electrical signals into audio (sound) signals. The signal generators that we discussed first create *electronic signals*; these devices generate *sound waves*. For the most part, the only types of sound generating equipment we commonly use are speakers. There is, however, a large variety of speakers, some of which do not look or operate like the box-type speakers we are so familiar with. Headphones are also properly considered sound generators.
- **Transmission media** are the methods we use to send signals from one place to another. Radio signals or other means can be used, but in almost every application, the choice is copper conductors. For small systems, the choice is almost always

two-wire speaker cable. It is essentially lamp cord (in some places called zip-cord) with clear insulation. Gauges typical run from 22 gauge (not recommended, except for the smallest of installations) to 16 gauge, for larger installations with longer runs. For critical applications, shielded twisted pair or coaxial cables are occasionally used.

Impedance Matching

Audio amplifiers can produce only a limited amount of current to drive speakers. If too many speakers are connected in parallel (which is the normal connection) to an amplifier, it will be overloaded and either function poorly or be damaged.

Audio amplifiers have their output limits measured in ohms. For example, an amplifier may have an output rating of 3 Ω (ohms). This indicates that the circuit you connect to the output terminals (the speaker connections) must have an impedance (total resistance, including inductive and capacitive reactance) of no less than 3 Ω.

Most speakers are rated at 8 Ω. Thus, two such speakers connected in parallel would give the circuits an impedance of 4 Ω (8 ohms per branch, divided by the number of equal branches). Three such speakers connected to a circuit would yield an impedance of 2.67 Ω (8 ohms divided by 3 branches). In these calculations we are ignoring the resistance value of the conductors, which is negligible except when very long runs are required.

Autotransformers

There are a few amplifiers that can handle loads with impedances as low as 1.5 Ω, but they are far from common. In most cases, if you wish to connect more than two speakers (8 ohms) on a circuit, you will have to install an autotransformer to adjust the impedance of multiple speaker pairs and present a sufficiently high impedance to the amplifier. You will often find these audio autotransformers sold as *impedance matching transformers*.

Volume controls are often autotransformers (they can also be potentiometers), and they can be used to reduce the level of signal sent to the speakers, while still presenting the same circuit impedance to the amplifier.

Components

Before we finish with the basics of sound systems, there are a few terms that should be defined:

- A **tuner** is an AM and FM receiver that is used to tune in a channel frequency and detect the station signal at that frequency. A tuner is frequently combined with an amplifier and enclosed in the same unit, which is often called a stereo receiver. Nonetheless, do not confuse the issue. A tuner cannot amplify the signal—only receive it. If you plugged speakers into a tuner (rather than a combination tuner/amp), they would not produce sound.

- An **amplifier (amp)** is a transistor-based signal amplification unit. Amps are used to take the low-power signals generated by tuners, turntables, and other detection equipment and raise them to high enough levels to run powerful speakers.
- A **preamplifier (preamp)** is used to boost a signal. Not as powerful as a normal amplifier, a preamp is necessary in some systems, or to drive certain speakers. It is normally used as an input to the main amplifier (power amplifier) that drives the speakers.
- An **equalizer** is a group of tuned circuits, used to color the sound produced by the system. By using an equalizer, it is possible to add to or diminish certain frequencies or frequency bands. For instance, if a sound system is producing a lot of hiss, using an equalizer to cut back on the high frequencies will eliminate it. There are two main types of equalizers: *graphic* and *parametric*. Graphic equalizers are the most common type, and likely the type you have seen; they modify bands of frequencies, such as from 30 to 800 Hz, from 1,000 to 5,000 Hz, and from 7,000 to 12,000 Hz. Parametric equalizers, on the other hand, boost or cut one specific frequency, or a very narrow band of frequencies. Graphic equalizers are most commonly used for home installations, and parametric equalizers are used for recording or special applications.
- A **tweeter** is a speaker specifically designed to reproduce high frequencies. It is usually used in combinations with other types of speakers.
- A **woofer** is a speaker specifically designed to reproduce low frequencies. It is usually used in combinations with other types of speakers.
- A **subwoofer** is a type of speaker used to reproduce very low frequencies, in the range of 30 to 125 Hz. Unlike normal speakers, subwoofers do not have to be aimed at the listener. Since such very low frequencies *propagate* (spread out) their sound over a very wide area, they can be placed in almost any part of a room with equally good effect. In other words, subwoofers tend to spill their sound, rather than radiating the sound in one direction only.

- The **RCA connector**, consisting of an **RCA plug** and an **RCA jack,** is the type of audio connector that is standard equipment on nearly every piece of stereo equipment sold. One conductor is connected to a central pin on the plug that fits into a sleeve in the jack, and the other conductor is connected to a concentric flange on the plug that surrounds the pin and fits over a ring on the jack.
- The **signal-to-noise** ratio is the generally used measure of signal quality for stereo systems. It is a comparison of the signal level to the noise level. It is expressed in decibels; the higher the better.
- The **decibel** (**dB**) is the unit of measure of signal strength for all types of communication and sound systems. The decibel scale is logarithmic. A difference of 20 dB between two signals means that one signal has 10 times the amplitude of the other; a difference of 40 dB means that one signal has 100 times the amplitude of the other. The decibel scale was originally developed for the telephone system. and the bel (10 decibels) is named after Alexander Graham Bell.
- A **phase-locked-loop,** or **PLL,** is a special type of tuning circuit that uses a quartz crystal vibrating at a specific reference frequency. It is used for digital tuning, and generally produces more accurate results than other methods.
- The amount of speaker-driving power that an amplifier can produce is measured in **watts per channel.** A stereo system has two channels.
- **Ambient noise** is the background noise in a room or area.
- **White noise** is a sound signal containing an even distribution of all frequency levels. White noise is used to test a room to determine its acoustical characteristics.
- **Dolby** technology is used for noise reduction in recording and playback in many sound systems. There are several types, including Dolby A, B, and C.
- **DBX** is another noise reduction system, similar in concept to Dolby, but it operates differently.

The Code

Sound wiring is subject to the requirements of *Article 725* of the *NEC.* In almost every possible case these rules call for nothing more than supporting the proper types of cables and for not connecting them to anything but special equipment.

As mentioned in Chapter 7, coaxial wiring is covered by *Article 820* of the *NEC*. The primary safety requirement of this article is that the voltage applied to coaxial cables cannot exceed 60 volts, and the power source must be energy-limited.

Coax cables indoors must be kept away from power or Class 1 circuits, unless the circuits are in a raceway, metal-sheathed cables, or UF cables.

Coaxial cables can be run in the same raceway (or enclosure) with Class 2 or 3 circuits, power-limited fire-protective signaling circuits, communications circuits, or optical cables. They may not be run in the same raceway or enclosure with Class 1 or power conductors. Exceptions are made if there are permanent dividers in the raceway or enclosure, or in junction boxes used solely as power feeds to the cables.

Coax cables are allowed to be run in the same shaft as power and class 1 conductors, but in these cases, they must remain at least 2 inches away. (This applies to open conductors, not to conductors in raceways, metal-sheathed cables, or UF cables.)

Installation

The installation methods for these systems are very basic and require little in the way of new tools. In some cases, a multimeter designed for very small current and voltage levels may be necessary for testing.

Acoustics

Acoustics is a general term referring to the study of the behavior and perception of sound. When designing a sound system, it is important to have a basic understanding of how sound behaves in a room, and how it is received by the human ear. A few of the more important points are as follows:

- Sound waves spread out from their source in the same way that waves on water spread out in a pond.
- Sound waves readily bounce off of hard surfaces such as plaster walls but are largely absorbed by soft flexible surfaces such as carpet or fabric.
- The angle at which sound waves bounce off of a hard surface is equal to, and opposite of, the angle at which they strike the surface.
- Sound waves that reach the ear from the side tend to have the most pleasing sound. (This can be difficult to implement well in the home environment.)

- Because sound travels relatively slowly, delay units are required in large or long structures. If they are not used, signals from distant speakers will reach the listener later than sound from nearby speakers, making the sounds jumbled and unappealing.

Speaker Placement

The first step in placing speakers is to define which rooms will be wired for sound.

The most common choices for high-quality stereo sound (sound produced by two different channels and speakers, giving a feeling of "separation") are game rooms, dens, master bedrooms, and dining rooms. Background sound is also desired in areas such as kitchens, hallways, bathrooms, and patios. These areas get monaural sound (one speaker for all the sound, which cannot deliver the separation effect).

Once you decide which rooms will get sound, you must spot the speakers in each room so that they produce the best effect. The general rules for placing speakers in rooms are as follows:

- Do not place speakers in the ceilings unless you have no other choice. Sound emitted downward will not easily fill a room well. Balance between the right and left channels can be heard poorly as well, depending on the listener's position in the room.
- Two stereo speakers should be placed with the distance between each of the two being equal to the distance between the listener and each speaker. For example, if your couch is seven feet from bookshelf-mounted speakers, the speakers should be mounted seven feet apart.
- Place in-wall speakers at about 5 feet above the finished floor (standing ear level). If the ceiling in the room is over ten feet high, mount the speaker at about half the height.
- Two-way speakers (those with separate woofers and tweeters) should generally be mounted so that the woofer (which reproduces the low-pitched sounds) is above the tweeter (which reproduces the high-pitched sounds).
- Make sure that you use weather-resistant speakers (sometimes called weatherized speakers) in areas such as garages and greenhouses.

Integrated Entertainment Systems

Virtually all types of electronic entertainment systems can be integrated into a home control system, giving the user far more flexibility and control than the entertainment devices would alone. The entertainment features most commonly tied into an automation scheme are sound systems. It seems that the owner or builder of almost every custom home is interested in some type of sound system.

Home Theater

Home theater systems are assemblies of television, video, and audio equipment connected together to deliver very high-quality entertainment programming. Systems like these go by several names—home theater, media systems, surround sound systems, and perhaps several other names. The most common components of such systems are the television receiver, possibly a cable or satellite TV converter, a videocassette player/recorder and/or DVD player, a stereo amplifier, and speakers. Optional equipment would include video games, radio receivers, cassette recorder/players, compact disc players, video distribution units, and computers.

The audio signal can either be sent to the television receiver's built-in speakers, or be sent through the stereo system, which produces a more powerful and clearer sound than the TV receiver. When special speakers and speaker placements are used, these systems are generically called surround sound, a term that properly refers to a trademarked brand of sound system approximating the sound quality of large motion picture theaters. Such systems make considerable use of subwoofer speakers to produce very powerful low-frequency sounds.

These systems are generally contained in a room dedicated to that purpose, with high-end equipment being used. Such complete systems typically add $5,000 to $15,000 to the cost of a home.

Surround processors provide six full channels of surround sound: front left, front center, front right, right surround, left surround, and low frequency, delivering the highest-quality sound. There should also be sufficient power to match the speakers you choose. As important, make certain that there is equal power to all six main channels.

Home Theater Terms and Components

There are several topics that should be covered before installing a home theater. Here are some definitions:

- The **audiovisual (A/V) receiver** plays the same role as stereo receivers, but the these devices include a built-in surround processor and several extra amplifier channels for connection to home theater speakers. A/V receivers should be capable of providing adequate power to your speakers and should include the capability to decode Dolby Digital (AC-3) signals. If you are using an A/V receiver, the only other components that you need to complete your system would be a VCR, a DVD player, a TV or video projection device, and quality speakers.

- **DBS (direct broadcast satellite)** systems allow you to access a wide variety of program materials that include movies, sports, music, news, information, educational, and pay-per-view programming. Available from vendors such as DirecTV and Primestar, these signals are bounced off of satellites and are received with small dish-type antennas, typically about 18 inches in diameter. To access these signals, you must lease or purchase the system and pay a monthly fee. The primary benefits of DBS is a much larger number of available channels.
- The **DVD (Digital Disc Versatile**, originally called the **Digital Video Disc**) is similar in shape and size to an audio compact disc and can contain a vast amount of information. When used for movies, the discs often contain the entire movie in several languages and several screen aspect ratios. DVDs can be purchased or rented, but you will need a special player to use them, as the discs contain digitally encoded information. Major benefits of DVDs include superior audio/video performance, instant access to chapters within a movie, and an increasingly vast library of titles.
- A **Hi-Fi VCR**, similar to a standard VCR, includes a special playback head that lets you access the high fidelity stereo audio tracks required for Dolby Surround and Dolby Pro-Logic analog Surround Sound playback. Hi-Fi VCRs are designed to connect with an A/V receiver, or surround processor's audio input and an associated video display device, provides a very convenient way to enjoy a home theater system. When selecting a tape for viewing, be sure to note whether your selection has been recorded for Dolby Surround. You should be able to find the Dolby logo on the container or the tape, which signifies that the movie is formatted for Surround Sound playback. Major benefits of a Hi-Fi VCR include an abundance of titles for rental or purchase and the addition of high fidelity Surround Sound audio tracks.
- **Laser Discs** were the first high-definition audio/video playback medium available for home theater systems. Approximately the size of vinyl long-playing phonograph records, these discs combine digital audio and video data and require a special laser

disc player for interface with the audio system and the video display device. Laser discs, although popular with early home theater enthusiasts, never achieved widespread acceptance in the sale and rental markets. Consumer acceptance has been low to moderate, but these discs offer movie lovers several key benefits, including the ability to view films in their original screen aspect ratios and to enjoy the audio soundtracks in Dolby Surround, Dolby Digital, and Digital Theater Sound (DTS) formats.

- **Line multipliers** also referred to as doublers, quadruplers, and scalers, are video processors that take standard-resolution video signals from VCRs, DVD players, DBS receivers, and even broadcast television signals and increase their resolution by a factor of two to four times the original. Because of the higher-frequency nature of the output signal and the way these processors separate the video signal into distinct colors and sync signals, they are used only with CRT, DLP, and rear projection sets that have sufficient frequency response.
- **Power amplifiers** are very simple devices in that they take a small signal and increase its power to reproduce sound through speakers. Home theater power amplifiers come in many configurations and power ranges. One amplifier channel is needed for each speaker in a home theater system. (Note that some subwoofers include their own internal power amplifier, so you might only require five channels of amplification for a complete home theater system.) Some power amplifiers combine all five or six amplifiers in a single chassis; others, which are sometimes called mono blocks, will power only a single speaker or pair of speakers. The size of the room, the efficiency of the speakers, and, ultimately, your individual taste will dictate your power requirements. Generally, amplifier power outputs of less than 50 watts per channel are considered low; 50 to 100 watts, medium; and over 100 watts, high.
- A **surround processor** is a stand-alone electronic component that would includes an audio/video input switching section, a Surround Sound decoding section, and an output control section. Unlike an A/V receiver, these units do not include an AM/FM tuner or amplifiers for the loudspeaker system. Dedicated surround processors are normally preferred for higher performance systems.

If You're Serious...

If your customer is really serious about a home theater, here are some recommendations:

- **A correctly sized screen.** When you go to the movies, the screen captivates your field of view, which is key to the immersing experience of a theater. If you choose not to fill the front of your home theater with a screen, be sure to select one large enough that you are not distracted by activities to the left and right of the screen. On the other hand, a screen that is too large for a room will make the scan lines in the picture too visible and will lessen the effects of a sharp, clean image. Selecting a screen with a picture width of half the distance from the seating position to the screen is close to ideal, though personal preferences vary. Provide for maximum input quality by insisting on S video connections for use where possible. The newest DVD players output their signals in an even higher quality form called Component Video. To get highest quality reproduction, your video display, be it a projector or direct view set, should have component video inputs, as well.
- **Surround Sound decoding and adequate amplifier power.** Sound is also a very important part of recreating the theater atmosphere. In fact, studies have shown that no matter how sharp and clear the picture is, the addition of a high-quality audio system not only makes programs sound better but also triggers emotional responses that actually make pictures look better. The best way to achieve this superior sound is to select an A/V receiver or surround processor that can decode the new digital audio systems and is compatible with tens of thousands of existing analog surround titles and television programs.
- **High-quality connections.** High-quality connection cables will aid in blocking out unwanted signals. They needn't be the most expensive wires; there are affordable interconnects and coaxial lead-in cables that protect the cables' signals from interference such as radio transmissions, cordless and cellular phones, microwave ovens, and nearby broadcast stations.
- **Matched speakers and subwoofers.** Having speakers that are matched at the three front channels (resulting in smooth pans across the front room) is essential to any home theater. If you are happy with your existing front left and front right speakers, look for a new center channel speaker that matches the

acoustic characteristics of the others. Surround speakers should also match one another, though they are typically smaller than the front speaker. To complete your speaker system, add a subwoofer. Most of the subwoofers available today are powered; that is, they include a built-in amplifier.

- **Lighting.** Often, lighting is treated as an afterthought, leading to a situation where the video image or the room ambience is not fully realized. On the other hand, if lighting is treated as an integral part of your home theater, the area will become a showcase for family and friends. Make every movie-watching event magical by dramatically dimming the lights at show time and gradually raising the lights again during the credits. Planning for a lighting system begins first by determining the effects, such as accent lighting, wall wash lighting, and task lighting, required for your room. Next, consider the type of lighting fixtures required. Depending upon the room décor and construction details, lighting fixture choices can include recessed downlights, track lighting, fluorescent fixtures, fiber optics, and emergency lighting. As a key visual element in the planning of your theater, a good lighting plan should provide the proper general light level, provide illumination at the perimeter surfaces, and highlight aesthetic elements of the room, such as artwork or interesting architectural details. This should be achieved while also striving to be as unobtrusive and energy efficient as possible.
- **Lighting controls.** We've come to expect, without even thinking about it, that when the lights dim, the movie is about to begin. Imagine yourself in a brightly lit movie theater. Just before show time, the lights are turned off abruptly, and you sit in total darkness. As the credits roll, the lights are thrown back on to full bright as your eyes try to recover from this sudden attack of light. It's hard on the eyes and certainly doesn't create a mood. To create movie theater lighting effects in your own home theater, replace toggle switches with dimmers, or perhaps a whole-room or whole-home lighting control system. Be certain that lighting is controlled in your viewing area so that there are no reflections on the screen and that room lighting does not otherwise distract from the screen images. A lighting control system gives you the ability to create and recall favorite lighting scenes, or preset light levels, with the touch of a button, even from the comfort of a favorite chair or bed. A number of scenes can be programmed for all of your home

theater activities: preshow, show time, intermission, credits, and cleanup. Then, recall the one you want with a single press of a button.

- **Placement.** To ensure that all the pieces of your new home theater fit together properly and deliver the performance you expect, it is important to do more than just set them down where your old TV and stereo were. Sound equipment requires careful placement, as well. Place the front speakers as far away from one another as they are from the viewing location, and put the center speaker directly above or below the viewing screen. Surround speakers should ideally be placed at the sides of the room, if possible, as they are in the movie theater, rather than on the rear wall.

Video Distribution

Video distribution systems allow for custom television channels to be transmitted through a home. This is generally done in combination with a cable TV system. Although these systems are capable of handling well over 100 stations, very few use more than 60 or 70. This allows the homeowner to transmit other signals through the unused channels. Channels are assigned with the use of a small electronic device called a *modulator*, which applies the basic television signal to any channel.

These extra channels can be used for any type of video signal at all, but the two most common uses are security monitoring of the baby's room, front yard, back yard, front and rear doors; and VCR or DVD signal distribution so that you can watch the movie playing on the VCR or DVD on any TV in the home.

Once there is a good television feed to a home, you can build a video distribution network within the home. This will almost always be included with a home theater system.

Designing a System

The design of a video distribution system is normally as follows:

1. Decide what signals are to be used: monitoring the doors, playing a DVD, and so forth.
2. Choose spare stations on which the signals will be placed. This will require you to identify stations that are not being used by the cable system that feeds the home. Care should also be taken to avoid stations that are likely to be put into service by the cable provider soon. (It might be a good idea to call the cable company and find out what they are planning.)

3. Choose a location for a small backboard (2 feet × 2 feet is usually fine) for mounting your equipment.

Necessary Equipment

The pieces of equipment that are normally used for video distribution are the following:

- **The modulator.** This is the essential piece of equipment that assigns raw video camera outputs to specific channel frequencies. It is a small electronic device. The system's coaxial cables simply plug into it.
- **Amplifiers.** These small electronic devices increase the strength of the video signal if it is not of sufficient strength. This is seldom a problem in an average sized home, but amplifiers can be necessary for larger homes or offices.
- **Splitters.** These operate as a Y connection, splitting the signal and sending it several different ways.
- **Surge suppressors.** These devices serve to keep power surges, spikes, and noise off the cables connecting the system. They are not required by the *Code* for interior work, but are fairly inexpensive, and are a fairly good idea.
- **Switchers.** These are video switches. They are used to put multiple inputs onto a single TV channel. Each of the various inputs may be used, not only one at a time, depending on the position of the switch.
- **Sequencers.** Sequencers put several inputs onto one channel, one at a time, in sequence.

Appendix D

The Internet as Entertainment

When properly connected (see Chapter 7), the Internet can become an alternative source of entertainment. Here are some basics about what you need to make the Internet work for you.

As most of us are aware by now, the Internet (or Net) is simply a group of connections between computer networks. It is a large group of high-speed communication lines (almost all of them optical fiber) that connect thousands of computer centers together. These communication lines are owned by AT&T, Sprint, MCI, UUNet, and several other companies, who rent access to these high-speed lines. These computer sites then connect to millions of individual homes and offices via regular or digital telephone lines.

No one really owns the Internet. Every piece of the Net is owned by someone, but that little piece is not necessary for the entire Net to operate. There are no charges on the Internet itself: no long-distance charges, no online charges, and no dues. The people who connect you to the Net, such as America Online (AOL), Prodigy, or local or regional providers, charge for *access* through their system to the Internet; and the owners of main Internet lines (backbones) charge for access to the network, but once you get into the Net, there are no further charges.

Furthermore, there are no preset paths for communications on the Net. There are no switches to connect your computer to some other computer. There is no central brain. When you send an Internet message, your computer begins the message with an electronic address. When your message reaches the next junction point, the machine there (called a router) reads the message, identifies the best path towards that destination, and sends it along. When your message reaches the next crossroad, the same thing happens. Your message may hit five or ten crossroads before it gets to its destination, and when your friend on the other end replies to your message, the reply will probably come back to you along a different path. But as long as the addresses are valid, the messages will get to where they belong. You can see from this why the Internet is called a routed network, whereas the telephone system is called a switched network.

Why Use a Computer for Entertainment?

Televisions now get hundreds of channels, there are probably dozens of radio stations in your area, and in all likelihood there are several multiplex theaters within driving range. So, why would someone need to use a computer for entertainment? There are actually several reasons:

- **Price:** Entertainment available over the Internet is free. People share it among themselves.
- **Variety:** In a big city, there are probably a couple of dozen radio stations available. On the Internet there are thousands. If you live in Toronto and like a talk station in Atlanta, you're out of luck—unless you have an Internet connection. If you do, you can listen to your station day and night. There are literally hundreds of talk stations, rock stations (hard rock, classic rock, light rock, romantic rock, and rock from specific periods), classical stations, jazz stations, news stations, and virtually any other type of radio station you can imagine. If you'd like to hear some specific pieces of music, they are all available instantly. Looking for an old Beethoven choral piece? Or an obscure pop hit from 1958? No problem—you can find it right away on the Internet.
- **Video programming:** It is not yet as available on the Net as is music (the files are much larger and are difficult to download without a broadband connection), but there's far more of it available than you might expect, and increasing quickly. There is no functional difference between storing or sending a music file or video file; only the size of the file differs.
- **Availability:** There's no running to the video store or the music store; just do a quick search and download the material you want. You never have to leave the house.
- **Customization:** With traditional entertainment, what a supplier gives you is all you get. TV and radio shows come full of commercials, ad there is little or nothing to be done about it. Songs and programs are available only at certain times, and then you may never hear or see them again. Not so with computer-based entertainment. Commercials can be removed. Audio can be modified, video can be edited (a surprising number of people are equipped to edit video), and many more custom modifications are sure to be appearing in the next few years.

It is hard not to notice that computers are becoming a central part of modern life. A large majority of households in the United States now have at least one, and most of Europe and much of Asia aren't far behind. The market for running entertainment through a computer is big now and will probably be the standard in a few years. This is the time when you have to get used to it. From this point on, a computer of some type will become the interface between the entire world of available programming and the residence.

Computers Outfitted for Entertainment

A computer-based entertainment setup allows you to enjoy every type of entertainment—TV, video, music, radio, CD and DVD, as well as downloading, recording, and playing all of your audio through a high-end stereo setup—all on a standard computer that can also be used for word processing, accounting, surfing the Net, exchanging e-mail, connecting to a network, or whatever else you would normally do.

First, here is a list of the various components required for the kind of arrangement under discussion:

1. Computer: processor, mouse, keyboard, NIC, monitor
2. Equipment case
3. Video card
4. MPEG decoder
5. Sound card
6. Hard drives—separate hard drives are recommended for video, for MP3, and for the operating system (OS) and applications
7. Optical storage (DVD, CD-RW, or combo device)
8. Universal Serial Bus (USB) hub
9. Video display (high resolution)
10. Speakers and receiver (amplifier?)
11. Software
12. Computer cables
13. Network cables (to the Internet)

The Computer

For a good PC-based entertainment system, the computer needs to be beefed-up just a bit more than usual. Obviously it should begin with a good processor. Processors improve seemingly every month so don't bother trying to get the very best and newest. Do, however, get a relatively new type of processor—not top of the line, but no more than two or three steps down from it. Processing speed is not really critical in any of these applications, but if your customer wants to run a lot of applications at the same time, the extra power will be helpful.

You will, as usual, need a mouse, keyboard, and monitor (get a good monitor—it is worth it). Also necessary will be a a network interface card (NIC), so that you can connect to some type of broadband Internet connection.

Note that item number six in the preceding list specifies three separate hard drives. It is not required that you use three separate hard drives, but doing so will make the system operate more efficiently over time. You do, however, need to use big hard drives. The list specifies one hard drive for the operating system (OS) and application programs (such as word processing, accounting, and so on), another for storing video files (40 GB would be ideal), and one more for storing music files (15 GB would be great). Again, you could use a single hard drive for all of this, but you'd have to get a large one. Fortunately, hard drive prices (especially dollar-per-storage capacity prices) have plummeted in the past several years, making these components easily affordable.

Finally, remember that computer components get bigger and better all the time. The items we are specifying here are good for the time of the writing, but not necessarily for next year. Choose good, modern components.

Equipment Case

The items just specified will not all fit easily into an off-the-shelf computer; there are simply too many of them, and average computers are not made with enough room for them all. This being the case, a system built with all these items will need a special equipment case, built into the computer. This means that you'll have to have someone custom-build the computer for you, rather than picking it off the shelf at CompUSA. This is not a problem, because the prices for custom-configured computers are not much higher than for the off-the-shelf models, although it will require a couple of phone calls and e-mails or faxes to get the information transferred from you to the builder.

The type of case you want is called an ATX case, and a good one now costs about $300. All the equipment specified should fit into the case, although there could be a little bit of modification involved. Generally, this requires nothing more than a pair of tin snips or a small Dremel tool.

Also bear in mind that your case will need a power supply. Some of them come with a built-in power supply (with cooling fan), and some do not. Make sure you have one.

Video Card

All standard desktop computers come with a video card. This card, however, is a special one. It does far more than a standard video card, and it will be something that you purchase in addition to the standard video card that comes built in to your computer.

You will need a separate video card that has a television tuner built in, an MPEG capture and editing feature, and perhaps a PVR feature. (*PVR* stands for Private Video Recorder. If you have seen commercials for TiVo, this does approximately the same things.)

The best video cards for this sell for about $300 currently. A good card can be purchased for a bit less than $200.

MPEG Decoder

MPEG is the file format developed by the Motion Pictures Engineers Guild. An MPEG file is a digital video file. These files are almost always compressed (to save transfer times and disk space), so they must be decoded. Also, the files must be decoded properly for the type of display you will be using.

The DVD Motion card (about $50) seems to be among the best for this job, and it will give you the ability to drive large, wide screens very well.

Sound Card

As with the video card, this is not the simple type of sound card that comes with a garden variety computer. You'll want a sound card with multiple digital inputs and analog outputs.

Right now, the best sound cards cost about $175. You could, however, use a lesser card, for about $35.

Optical Storage

Optical storage is a device that can read and write to an optical disk such as a CD or DVD. A CD burner is an optical storage device.

Instead of buying several stand-alone devices (DVD player, CD burner), it is generally better to buy a combination drive, which sells for under $300. Buying the combo device makes operating the system much easier, and the cost is pretty close to the same.

USB Hub

The USB hub is required to extend your computer's existing serial bus and to allow more devices to connect to it. The serial bus is one of your computer's main internal means of communication. Any type of USB hub will work for this application, provided that it is relatively easy to mount.

Video Display

Here's where it gets interesting, and potentially expensive. The TV that sits in your living room is actually a very low-resolution device. Even a rather cheap video monitor will give significantly better video quality. And since we already have a television tuner built in to the video card we specified, there is no necessity of a second one in a television set. You don't need to have a complete TV set—only a nice monitor.

You do not have to stick with the standard TV-screen shape (a 4 : 3 aspect ratio). Instead, you can use a wide screen (16 : 9 aspect ratio). This is much nicer for viewing movies and is no worse for viewing TV programming. HDTV signals are being broadcast in this wide perspective also.

To get a really great video monitor you could pay $15,000. For a good but not great display, you could pay about $2,000. For an inexpensive 21-inch monitor, you would pay in the range of six or seven hundred dollars. All of these will be substantially better displays than any standard television. The display is critically important; if your customer wants to watch any type of video production, the extra money is well worth it.

There is one risky area here: New types of video monitors are being developed all the time, and the various input arrangements on them vary. Make sure that the monitor you choose will match the system you are building. It would be a good idea to take a photo, or make a good drawing, of the equipment you intend to connect the monitor to, then be sure that the inputs and outputs match. Test the monitor as soon as you get it on site, and return it immediately if it doesn't work with your system. In several years all of the compatibility issues should be worked out, but for the meantime, you'll have to pay some attention to this.

Speakers and Receiver

The big question here is "How serious are you about good sound?" With modern stereo equipment, very good sound is available at a reasonable price; but if your client wants great sound, the price can go up and up and up. Another question is how much of a theaterlike experience you want to achieve.

The basic setup for most installations (both high-end and low) is for five speakers plus a subwoofer, as described in Appendix C. Obviously this must be coupled with a receiver.

If you are interested in high-quality sound, you absolutely need equipment that is capable of decoding digital signals. All the soundtracks on DVDs are digital, so this really is a necessity. There are two digital sound formats for home theater systems: Dolby Digital (DD) or Digital Theater Sound (DTS). Dolby Digital 5.1 is the standard for DVDs.

The types of systems available range from an okay $300 system to a terrific $3,000 system, and as far up from there as you'd like to go. (The prices really do get ridiculous.)

Software

There are actually three pieces of software you need for a good system: software for MP3 sound files, PVR software (probably comes with the equipment), and DVD software (which should also come with the equipment).

Cables

You'll need a variety of cables to connect all of this equipment. You don't need the gold-plated type (and, yes, they are sold in gold-plated versions), but do get good-quality cables.

The Remote

You'll want a lot of your equipment to be controlled with some type of remote control. But since there's a lot of it, you don't want several remotes to keep track of. You can solve this problem by purchasing a programmable universal remote.

Index

C